AI, Cybersecurity and Data Science for Drone and Unmanned Aerial Vehicles

This book explores the transformative impact of drone technology and unmanned aerial systems (UAS) across diverse industries, from precision agriculture and logistics to disaster response and forensic investigations. It highlights how the integration of Artificial Intelligence (AI) into UAVs is addressing contemporary challenges, optimizing operations, and shaping the future of aerial systems. It offers technical methodologies, case studies, and a detailed roadmap for integrated drone forensics, making it an essential resource for understanding the dynamic sector of *Intelligent Aerial Systems*.

- Examines the integration of AI in drones and UAS, enhancing operational efficiency and investigative processes
- Covers the evolving landscape of drone technology, including swarms, autonomous systems, and forensic methodologies
- Focuses on specific sectors such as agriculture, logistics, and disaster response, offering practical insights, real-life examples, and future scenarios
- Explores the role of drones in delivery and logistics, transforming transportation systems by improving efficiency and reducing costs
- Explains the use of drones for public safety, smart city disaster response, and forensic investigations, providing critical insights for emergency operations

This reference book serves as a comprehensive guide for professionals, researchers, and scholars interested in the rapidly advancing world of drone technology.

AI, Cybersecurity and Data Science for Drone and Unmanned Aerial Vehicles

Real-Life Applications and Case Studies

Edited by
Shishir Kumar Shandilya,
Fernando Ortiz-Rodriguez,
Smita Shandilya and Gerardo Romero

CRC Press
Taylor & Francis Group
Boca Raton London New York

CRC Press is an imprint of the
Taylor & Francis Group, an **informa** business
A CHAPMAN & HALL BOOK

Designed cover image: Shutterstock

First edition published 2026
by CRC Press
2385 NW Executive Center Drive, Suite 320, Boca Raton FL 33431

and by CRC Press
4 Park Square, Milton Park, Abingdon, Oxon, OX14 4RN

CRC Press is an imprint of Taylor & Francis Group, LLC

ISBN: 9781032766140 (hbk)
ISBN: 9781032771588 (pbk)
ISBN: 9781003481584 (ebk)

DOI: 10.1201/9781003481584

Typeset in Times
by Deanta Global Publishing Services, Chennai, India

Contents

Acknowledgments

We unequivocally believe in being thankful through action, not just by saying "thank you" or merely acknowledging support. We are fortunate that our journey into the fascinating fields of AI, cybersecurity, and data science for UAVs has been supported by individuals whose contributions were invaluable to this book.

We are deeply grateful to Aditya Srivastav for his exceptional assistance throughout this project. His technical expertise, attention to detail, and innovative approach provided critical support during the development of this book. His dedication and problem-solving skills have significantly contributed to the quality and depth of the content.

We wish to express our deepest gratitude to all the contributors of this book. Each of you, with your unique research and perspectives, has played an essential role in creating a comprehensive work that bridges theory and real-life applications. Your dedication to advancing the fields of AI, cybersecurity, and UAV technology has made this book a valuable resource for both academia and industry.

We also extend our sincere thanks to our families, friends, and colleagues for their patience, support, and encouragement throughout this journey. Your understanding and words of motivation have been a constant source of strength. We are grateful to our publishers, CRC Press – Taylor & Francis UK, and the entire editorial team who worked diligently behind the scenes. Your professionalism, commitment, and attention to detail have been critical in bringing this book to life.

Lastly, but by no means least, we would like to acknowledge our readers. We hope this book serves as a practical and insightful tool in your exploration of AI, cybersecurity, and UAV technologies. We look forward to your feedback and engagement as we continue to advance in these fields.

Indore, India – Dr. Shishir Kumar Shandilya

Mexico – Dr. Fernando Ortiz-Rodriguez

Indore, India – Dr. Smita Shandilya

Mexico – Dr. Gerardo Romero

Editor Biographies

SHISHIR KUMAR SHANDILYA

Shishir Kumar Shandilya is an Associate Professor in the School of Data Science and Forecasting at Devi Ahilya University in India. He is also a Visiting Professor at Liverpool Hope University in the United Kingdom. He is a Cambridge University-certified professional teacher and trainer, a TEDx speaker, an ACM Distinguished Speaker, and a senior member of IEEE. Dr. Shandilya is a NASSCOM-certified master trainer for security analysts in SOC (SSC/Q0909: 7 NVEQF Level 7). He has received the IDA Teaching Excellence Award for distinctive use of technology in teaching by the Indian Didactics Association in Bangalore (2016) and the Young Scientist Award for two consecutive years, 2005 and 2006, by the MP Science Congress and MP Council of Science and Technology. He is a highly regarded author with published research works in reputable academic publishers. Dr. Shandilya also has international and national patents and copyrights granted for adaptive cyber defense methods. His research interests include adaptive defense, advanced digital investigation, and forensic methods, explainable artificial intelligence, privacy-preserving computing, nature-inspired cryptography (NIC), and nature-inspired cyber security (NICS).

FERNANDO ORTIZ-RODRIGUEZ

At present, Fernando Ortiz-Rodriguez is working at the Tamaulipas Autonomous University (UAT). He is a full professor and the Artificial Intelligence and Innovation (AI&I) Lab Founder and Head, corporate trainer, business and AI coach, and Information Technology strategic advisor for the National Manufacturing Association (INDEX). He is a member of level one of national systems researchers (SNI) of the prestigious National Council for Science and Technology (Conacyt) and member of the Association for Computing Machinery (ACM). He received 2021 and 2022 Latin American awards (u-gob) for implementing and developing software to help with COVID-19 diagnosis in Mexico.

SMITA SHANDILYA

Smita Shandilya (Senior Member-IEEE) is a Professor at SAGE University, Indore, with extensive experience in teaching and research. She has delivered numerous invited talks at national seminars and has published research papers in reputed international and national journals and conferences. Her primary research interests include Artificial Intelligence and Data Science. She is an integral part of the research and development section at her institute and has contributed to multiple research projects.

GERARDO ROMERO

Gerardo Romero was born in Zacatecas, Nuevo Leon, Mexico. He received the B.S., M.S., and Ph.D. degrees in automatic control from the Autonomous University of Nuevo Leon, Mexico, in 1990, 1993, and 1997, respectively. He has been the Head of the Department of Electronics from 1999 to 2014 and the Head of Graduate Studies and Research since 2014, at the UAM Reynosa Rodhe of the Autonomous University of Tamaulipas. His research interests include robust stability analysis and control of linear and non-linear systems and their applications to robots and industrial processes.

Contributors

Ramesh A
Sri Eshwar College of Engineering
India

Dhanya Y Bharadwaj
CHRIST University
India

Sudharsan K K
Sri Eshwar College of Engineering
India

Christian Kaunert
Dublin City University
Ireland & University of South Wales UK

Fernando Ortiz-Rodriguez
Universidad Autónoma de Tamaulipas
Mexico

Sivakumar P
PSG College of Technology
India

Srilakshmi Prabhu
Cambridge Institute of Technology North Campus
India

Roheen Qamar
Quaid-e-Awam University
Pakistan

Sri Sathya Priya R
Sri Eshwar College of Engineering
India

Velmurugan S
Sri Eshwar College of Engineering
India

Shishir Kumar Shandilya
Devi Ahilya Vishwavidyalaya
India

Smita Shandilya
Devi Ahilya Vishwavidyalaya
India

Bhupinder Singh
Sharda University
India

Aditya Srivastav
VIT Bhopal University
India

Devangana Sujay
VIT Bhopal University
India

Ruthresh T
Sri Eshwar College of Engineering
India

Baqar Ali Zardari
Quaid-e-Awam University
Pakistan

1 Introduction to Drones, UAVs, and Their Applications

Aditya Srivastav, Smita Shandilya, and Fernando Ortiz-Rodriguez

1.1 INTRODUCTION

1.1.1 Definition of Drones and UAVs

Unmanned Aerial Vehicles (UAVs), commonly referred to as drones, are aircraft that function without a human pilot present. With the use of inbuilt sensors and pre-programmed flight plans, these gadgets can fly independently or be remotely controlled by an operator. The term "drone" often encompasses various types of UAVs that serve different purposes, ranging from military applications to commercial and recreational uses. UAVs are integral to modern technology due to their versatility and wide range of applications, including military, construction, medical, search and rescue, parcel delivery, and more (Mohsan et al., 2022). UAVs belong to a larger group called unmanned aircraft systems (UAS), which consists of the UAV, a controller located on the ground, and a communication system connecting the two. These systems can vary significantly in size, design, and capability, depending on their intended use. For example, small consumer drones used for aerial photography differ greatly from large military drones used for reconnaissance and combat missions.

A number of criteria, including size, weight, design, and purpose, are used to categorize drones. They fall into three general categories: hybrid drones, rotary drones, and fixed-wing drones. Similar to conventional airplanes, fixed-wing drones are renowned for their extended flying times and capacity to cover wide regions. Because they can hover in place and are more nimble, rotary-wing drones—such as quadcopters and hexacopters—are perfect for tasks like aerial photography and inspection. In order to maximize performance for particular jobs, hybrid drones use components of both fixed-wing and rotary-wing designs (Otto et al., 2018).

Drones or UAVs are unmanned aircraft controlled either remotely or autonomously. They encompass a wide range of designs and applications, from small consumer drones to large military UAVs, and are a crucial part of modern technological advancements.

DOI: 10.1201/9781003481584-1

1.1.2 Historical Background of Drones and UAVs

The concept of unmanned flight has a rich history that dates back to the early 20th century. The first significant step towards modern UAVs was the development of the "Kettering Bug," an early cruise missile developed by the United States during World War I. Although it never saw combat, it laid the foundation for future UAV technologies (Sullivan, 2005).

During World War II, both the Allies and Axis powers experimented with various forms of unmanned aircraft. The development of radio-controlled target drones, such as the Radioplane OQ-2, provided essential training for anti-aircraft gunners. These early drones were primarily used for target practice but demonstrated the potential for more advanced applications (Valavanis & Kontitsis, 2007).

Due to the necessity of information and reconnaissance, UAV technology advanced significantly during the Cold War. During the Vietnam War, the United States produced a number of high-altitude reconnaissance drones, including the Ryan Model 147, which were widely employed for observation tasks. These drones could fly at high altitudes and capture detailed imagery of enemy positions, providing critical intelligence without risking pilot lives (Ehrhard, 2010).

In the late 20th and early 21st centuries, UAV technology experienced rapid growth, spurred by advancements in electronics, materials, and navigation systems. The development of the Predator drone in the 1990s marked a significant milestone, as it combined long-endurance flight capabilities with real-time video surveillance. The Predator's success in military operations, particularly in Afghanistan and Iraq, highlighted the strategic advantages of UAVs in modern warfare (Kindervater, 2016).

Today, drones are used in a wide range of applications beyond military purposes. In agriculture, drones facilitate precision farming by providing detailed aerial surveys and optimizing resource use. In construction and real estate, they offer significant benefits in site surveying, infrastructure inspection, and marketing through aerial photography. The logistics industry has also adopted drones for faster and more efficient delivery services, particularly in remote or difficult-to-access areas (Chiang et al., 2019).

1.1.3 Importance and Relevance of Drones in Modern Technology

Drones have become indispensable in various industries, significantly enhancing productivity, safety, and efficiency. Their importance and relevance in modern technology are evident across multiple sectors, including agriculture, construction, logistics, and emergency response.

In agriculture, drones are revolutionizing farming practices through precision agriculture. They offer farmers precise aerial imagery to evaluate field conditions, track crop health, and maximize fertilization and irrigation. This data-driven strategy contributes to more sustainable agricultural methods by increasing crop yields and decreasing resource waste (Tkáč & Mésároš, 2019).

The construction and real estate industries also benefit greatly from drone technology. Drones provide an economical and effective means of surveying building sites, tracking developments, and examining infrastructure. Better planning and decision-making are made possible by the high-resolution photos and 3D models they offer. In real estate, drones enhance marketing efforts by capturing stunning aerial views of properties, attracting potential buyers and investors (Otto et al., 2018).

In logistics, drones are transforming delivery services by providing faster and more reliable transportation of goods. They are particularly useful in last-mile delivery, reaching remote or hard-to-access areas where traditional delivery methods are inefficient. Companies like Amazon and UPS are exploring drone delivery solutions to enhance their logistics networks and improve customer satisfaction (Chiang et al., 2019).

Emergency response and disaster management are other critical areas where drones have proven invaluable. In the aftermath of natural disasters, drones can deliver necessary supplies, find survivors, and assess damage rapidly. They give emergency responders access to real-time aerial imagery, which aids in resource allocation and decision-making. Additionally, drones are employed in search and rescue efforts to increase their effectiveness and security (Bhatt et al., 2018).

Moreover, drones are making significant contributions to environmental monitoring and conservation efforts. They are used to track wildlife, monitor deforestation, and assess the health of ecosystems. Drones provide valuable data that supports conservation initiatives and helps mitigate environmental damage (Tkáč & Mésároš, 2019).

1.2 TYPES OF DRONES AND UAVS

1.2.1 Classification by Size and Weight

Drones, also known as Unmanned Aerial Vehicles (UAVs), come in various sizes and weights, which determine their applications and capabilities. The classification by size and weight can be broadly divided into three categories: Nano and Micro Drones, Mini and Small Drones, and Medium and Large Drones.

1.2.1.1 Nano and Micro Drones

Nano and micro drones are the smallest UAVs, typically weighing less than 250 grams. These drones are often used for indoor applications, educational purposes, and hobbyist activities due to their small size and ease of maneuverability. Despite their tiny size, these drones can be equipped with cameras and sensors for simple tasks like aerial photography and short-range surveillance. The primary advantage of nano and micro drones is their ability to operate in confined spaces where larger drones cannot fit. However, their small size limits their battery life and payload capacity, making them unsuitable for long-duration missions (Hassanalian & Abdelkefi, 2017).

1.2.1.2 Mini Drones

Mini drones, weighing between 250 grams and 2 kilograms, bridge the gap between the smallest drones and more capable small drones. They offer enhanced features suitable for enthusiasts and professionals, including improved flight stability, longer battery life, and better payload capacities compared to nano and micro drones. Mini drones are commonly used for aerial photography, videography, and small-scale surveying. They offer advanced camera systems, intelligent flight modes, and improved stability, making them versatile tools in various industries (Tsach et al., 2010).

1.2.1.3 Small Drones

Small drones, weighing between 2 kilograms and 25 kilograms, are designed for professional applications requiring higher payload capacities and advanced functionalities. They are used in search and rescue operations, precision agriculture, mapping, industrial inspections, and professional filmmaking. With the ability to carry specialized sensors and equipment, these drones can fly for longer periods of time and remain more stable in difficult conditions. Their greater payload capacity makes it possible to incorporate top-notch cameras, Light Detection and Ranging (LIDAR) systems, and other cutting-edge sensors, increasing their usefulness in a variety of industries (Elmeseiry et al., 2021).

1.2.1.4 Medium and Large Drones

Medium drones, weighing from 25 kilograms to 150 kilograms, are utilized in industrial and commercial operations that require heavy payloads and extended flight times. Typical applications include agricultural spraying, heavy payload delivery, large-scale surveying, and environmental monitoring. They are essential in sectors like agriculture, where they efficiently manage large fields, and in industrial inspection, where they access hard-to-reach areas. Large drones, weighing between 150 kilograms and 600 kilograms, are often used for specialized commercial and military applications. These drones are capable of long-endurance flights and can carry sophisticated sensor packages. Applications include surveillance and reconnaissance, border patrol, maritime surveillance, and long-endurance missions (Elmeseiry et al., 2021).

1.2.1.5 Very Large Drones

Very large drones, weighing over 600 kilograms, are advanced unmanned aircraft used primarily in military operations for combat and strategic reconnaissance. They can carry significant payloads, including weapons and high-end sensors, and operate at high altitudes for extended durations. Applications include military combat, strategic surveillance, and high-altitude long-endurance missions, providing critical intelligence and operational capabilities. Examples include drones like the General Atomics MQ-9 Reaper and Northrop Grumman RQ-4 Global Hawk, which have played significant roles in modern military operations, demonstrating the capabilities and impact of very large drones in defense strategies.

1.2.2 Classification by Usage

Drones can also be classified based on their intended use, which includes consumer, commercial, industrial, and military applications. Each category has specific requirements and capabilities that suit different operational needs.

1.2.2.1 Consumer Drones

Consumer drones are primarily designed for personal use and recreational activities. They are typically small, easy to operate, and affordable, making them accessible to a wide range of users.

1.2.2.1.1 Recreational Use

Hobbyist drones (also known as Recreational UAVs) are used for enjoyment and have the ability to fly over short distances with minimal restrictions. The entry-level drones usually come with very basic cameras and are able to perform standard airborne maneuvers. They are good for first-time users because of their simplicity. The recent high demand for drones with recreational use has also increased the interest in drone racing which is a sport where players fly around obstacle courses at high speeds. But besides those, there arc recreational drones that people can use to capture personal memories from diverse aerial perspectives which could be a new way of documenting events and travel experiences.

1.2.2.1.2 Aerial Photography

Drones for aerial photography are equipped with high-quality cameras capable of capturing stunning aerial images and videos. They are popular among photographers and videographers for capturing unique perspectives and creating professional-quality content. Consumer drones like the DJI Mavic series offer advanced features such as 4K video recording, gimbal stabilization, and automated flight modes, making them powerful tools for both amateur and professional photographers (Hildebrand,

Drone Size	Weight Range[1]	Example Drones	Typical Applications
Nano and Micro Drones	Under 250 grams	DJI Mini 3 Pro, Parrot Anafi, Hubsan X4	Recreational flying, indoor use, hobbyist photography, education
Mini Drones	250 grams – 2 kg	DJI Mavic 3, Autel EVO II, Skydio 2	Aerial photography, videography, surveying, real estate, media production, small-scale mapping
Small Drones	2 kg – 25 kg	DJI Matrice 300 RTK, Freefly Alta 8, Yuneec H520	Professional cinematography, industrial inspection, mapping, agriculture, search and rescue
Medium Drones	25 kg – 150 kg	DJI Agras T40, Yamaha RMAX	Agricultural spraying, heavy payload delivery, large-scale surveying, environmental monitoring
Large Drones	150 kg – 600 kg	Schiebel Camcopter S-100	Military surveillance and reconnaissance, border patrol, maritime surveillance, long-endurance missions
Very Large Drones	Over 600 kg	General Atomics MQ-9 Reaper, Northrop Grumman RQ-4 Global Hawk	Military combat and surveillance, strategic reconnaissance, high-altitude long-endurance missions

FIGURE 1.1 Drone size comparison chart[1]

Usage Category	Typical Applications	Description[1]
Consumer	Recreational flying, aerial photography, videography, drone racing	Drones used by hobbyists and enthusiasts for personal enjoyment and creative pursuits. These drones are generally affordable, easy to operate, and come equipped with basic features suitable for beginners and casual users. They often include cameras for capturing photos and videos from unique aerial perspectives.
Commercial	Agriculture monitoring, real estate photography, construction site mapping, media production, delivery services	Drones employed by businesses for professional purposes, offering advanced capabilities such as high-resolution imaging, precision navigation, and specialized sensors. They enhance efficiency and productivity in various industries by performing tasks like crop monitoring, surveying, aerial inspections, and even package delivery.
Industrial	Infrastructure inspection, energy sector monitoring, mining surveys, logistics and cargo transport	Robust drones designed for demanding industrial applications, capable of carrying heavy payloads, operating in harsh environments, and performing complex tasks. They are equipped with advanced technologies like LIDAR, thermal imaging, and long-range communication systems, supporting critical operations in sectors like energy, mining, and logistics.
Military	Surveillance, reconnaissance, combat missions, strategic intelligence gathering	Unmanned aerial vehicles utilized by defense forces for a range of military operations. These drones feature sophisticated systems for long-endurance missions, high-altitude flights, and can be equipped with weapons or advanced surveillance equipment. They play a crucial role in modern warfare by providing real-time intelligence and precision strike capabilities.

FIGURE 1.2 Drone usage comparison table[2]

2021). These drones enable users to capture breathtaking landscapes, dynamic sports action, and intricate architectural details from the air, significantly enhancing the creative possibilities in photography and videography.

1.2.2.2 Commercial Drones

Commercial drones are used in various industries to enhance productivity, efficiency, and safety. They are equipped with advanced technologies to perform specific tasks required in different sectors.

1.2.2.2.1 Agriculture

In agriculture, drones are used for precision farming, which involves monitoring crop health, assessing field conditions, and optimizing irrigation and fertilization. They help farmers make data-driven decisions, leading to better crop yields and resource management (Dutta & Goswami, 2020). Agricultural drones can perform tasks such as spraying pesticides and fertilizers, mapping fields, and analyzing soil health, reducing labor costs and increasing efficiency. For instance, drones equipped with multispectral sensors can detect crop stress and nutrient deficiencies, enabling targeted interventions that improve crop health and productivity (Bharti et al., 2020).

1.2.2.2.2 Real Estate

Aerial views of properties by drones are for real estate marketing and site inspection. Buyers appreciate these images because they allow a different view and help real estate agents show off properties much better. Aerial views captured by drones show off property features and landscapes (and even potentially your neighborhood!) in ways that traditional ground-level photography can't provide (Cvitanić, 2020). Drones can be employed to examine the roof and structure, picking up any issues that may need rectifying before the property changes hands.

1.2.2.3 Industrial Drones

Industrial drones are designed for heavy-duty applications and can operate in harsh environments. They are used for infrastructure inspection, energy sector applications, and more.

1.2.2.3.1 Infrastructure Inspection

Drones are used to examine vital infrastructure, including electrical lines, pipelines, and bridges. By offering high-resolution pictures and information, they lessen the need for human inspections and increase security. For instance, drones can inspect difficult-to-reach areas on tall structures, minimizing the risk to human inspectors and reducing downtime (Vohra et al., 2023). This capability is particularly beneficial for regular maintenance and emergency assessments following natural disasters or accidents.

1.2.2.3.2 Energy Sector

In the energy sector, drones inspect wind turbines, solar panels, and oil rigs. They help identify potential issues and perform maintenance tasks more efficiently, reducing downtime and operational costs. For example, drones equipped with thermal cameras can detect hot spots in solar panels, indicating areas that may be malfunctioning and require maintenance (Tkáč & Mésároš, 2019). Drones can also be used to inspect underwater structures in offshore oil rigs, providing valuable data without the need for divers.

1.2.2.4 Military Drones

Military drones are used for defense and combat purposes. They are equipped with advanced technologies for surveillance, reconnaissance, and strike missions.

1.2.2.4.1 Surveillance and Reconnaissance

Military drones are employed for combat monitoring and intelligence collection. They improve situational awareness and decision-making by offering real-time data and pictures. Drones such as the MQ-9 Reaper can loiter over areas of interest for extended periods, capturing high-resolution images and video that are crucial for identifying enemy positions and movements (Tsach et al., 2010).

1.2.2.4.2 Combat and Strike Missions

Combat drones are used for targeted strikes and combat operations. They can carry various payloads, including missiles and bombs, to engage enemy targets while minimizing risks to personnel. These drones can be remotely operated from safe locations, allowing military forces to conduct precision strikes with minimal collateral damage (Franke, 2014). The use of combat drones has transformed modern warfare by providing a platform for executing high-risk missions without exposing pilots to danger.

1.2.3 Classification by Design

Based on their design, drones can be divided into three categories: hybrid, rotary, and fixed-wing.

1.2.3.1 Fixed-Wing Drones

Drones with fixed wings look like regular airplanes and have lift-producing wings. They are renowned for their extended flying times and wide-ranging coverage.

1.2.3.1.1 Design and Features

Drones with fixed wings may glide in the air because of their stiff wing structure, which generates lift. To take off and land, they need a runway or a catapult. Because of their low energy consumption, these drones are appropriate for extended missions. Compared to rotary-wing drones, fixed-wing drones can fly higher and cover more ground, which makes them perfect for tasks like mapping and environmental monitoring (Garg, 2022).

1.2.3.1.2 Use Cases

Applications for fixed-wing drones include mapping, environmental monitoring, and surveying. They are perfect for applications requiring substantial data collecting over big territories because of their capacity to cover large areas. For example, fixed-wing drones are frequently employed in environmental conservation to study wildlife populations and habitats and in agriculture to check crop health (Ridwan & Alfindo, 2019).

1.2.3.2 Rotary-Wing Drones

Because they can hover in place and are more maneuverable, rotary-wing drones—such as quadcopters and hexacopters—are perfect for applications that need stability and accuracy.

1.2.3.2.1 Quadcopters

Quadcopters are the most common type of rotary-wing drones. They have four rotors that provide lift and stability, allowing them to hover, take off, and land vertically. They are popular for aerial photography and inspection tasks due to their ease of control and stability (Garg, 2022). Quadcopters are versatile and can be used in various applications, from consumer-grade recreational use to professional-grade inspections and surveys.

1.2.3.2.2 Hexacopters

Hexacopters have six rotors, providing greater stability and lifting capacity compared to quadcopters. They are used for more demanding tasks that require carrying heavier payloads, such as professional-grade cameras and sensors. The additional rotors provide redundancy, making hexacopters more reliable and less susceptible to failure if one rotor malfunctions (Hassanalian & Abdelkefi, 2017).

1.2.3.3 Hybrid Drones

Hybrid drones are engineered by combining elements of both fixed-wing and rotary-wing designs to optimize performance for specific tasks.

1.2.3.3.1 Design and Features

With the ability to take off and land vertically like rotary-wing drones and fly with the long-range efficiency of fixed-wing drones, hybrid drones combine both wings and rotors. They are adaptable and appropriate for a variety of uses due to their design. The ability to switch between hovering and forward flight modes gives hybrid drones the adaptability required for challenging operations (Okulski & Ławryńczuk, 2022).

1.2.3.3.2 Use Cases

Hybrid drones are used in applications requiring long-range flight with vertical takeoff and landing capabilities. They are ideal for missions such as search and rescue, where both agility and endurance are crucial. For example, hybrid drones can be used in disaster response scenarios to deliver supplies to inaccessible areas and perform search and rescue operations in challenging terrains (Krishnaraj et al., 2021).

1.2.4 Key Differences Between Types of Drones

The key differences between the types of drones lie in their design, capabilities, and applications. Understanding these distinctions is essential for selecting the appropriate drone for specific missions and for appreciating the technological diversity within the field of unmanned aerial vehicles (UAVs).

1.2.4.1 Advantages and Disadvantages

Each type of drone possesses inherent advantages and disadvantages that influence its suitability for particular tasks. Fixed-wing drones are renowned for their efficiency in long-distance and high-altitude flights due to their aerodynamic design, which allows them to glide on air currents and consume less energy once at cruising altitude. This efficiency translates into extended flight times and the ability to cover vast areas without the need for frequent refueling or battery replacement. However, fixed-wing drones typically require runways or catapult systems for takeoff and landing, limiting their operational flexibility, especially in constrained environments. Their inability to hover or perform vertical takeoffs and landings restricts their use in missions that require stationary observation or operation in confined spaces.

Rotary-wing drones, such as quadcopters and helicopters, offer exceptional maneuverability and the capability for vertical takeoff and landing (VTOL), making them ideal for operations in urban environments or areas with limited space. They can hover precisely over a target, ascend and descend vertically, and navigate in tight spaces, which is invaluable for tasks like aerial photography, infrastructure inspection, and search and rescue missions. The primary disadvantage of rotary-wing drones is their limited flight endurance. The energy required to maintain lift through constantly spinning rotors results in shorter flight times and reduced range compared to fixed-wing counterparts.

Hybrid drones are engineered with the aim to merge together the advantages of both fixed-wing and rotary-wing designs by incorporating features that allow for efficient forward flight and VTOL capabilities. These drones can transition between vertical and horizontal flight modes, enabling them to take off and land without runways while achieving greater speeds and longer ranges during cruise flight. However, the complexity of hybrid drones is significantly higher due to the integration of multiple propulsion and control systems. This complexity can lead to increased costs in design, production, and maintenance. Additionally, the weight penalties associated with combining different systems may affect payload capacity and overall efficiency (Elmeseiry et al., 2021).

Understanding these advantages and disadvantages is crucial for mission planning and drone selection. Operators must consider factors such as flight duration, range, payload requirements, operational environment, and logistical support when choosing the appropriate drone type for their specific needs.

1.2.4.2 Common Applications

The applications of drones are as varied as their designs, with each type lending itself to particular uses based on its capabilities. Fixed-wing drones are commonly employed in tasks that require covering large geographical areas efficiently. In surveying and mapping, they can capture high-resolution aerial imagery over expansive regions, facilitating the creation of detailed topographical maps and models. Environmental monitoring projects benefit from the extended flight times of fixed-wing drones, allowing for the collection of data on climate patterns, wildlife populations, and ecological changes over prolonged periods.

Rotary-wing drones excel in applications that demand precision, maneuverability, and the ability to hover. They are ideal for aerial photography and videography, providing stable platforms for capturing high-quality images and footage from various angles and elevations. In the field of inspection, rotary-wing drones can closely examine structures such as bridges, power lines, and wind turbines, identifying defects or areas requiring maintenance without the need for human workers to access potentially hazardous locations. Short-range surveillance and security operations leverage the agility of rotary-wing drones to monitor areas of interest, track movements, and respond rapidly to emerging situations.

Hybrid drones are particularly suited for complex missions that require a combination of long-range flight and VTOL capabilities. For instance, in logistics and delivery services, hybrid drones can transport goods over considerable distances and deliver them directly to locations without the infrastructure for traditional aircraft landings. In disaster response scenarios, they can quickly reach affected areas, delivering supplies, conducting reconnaissance, and providing communication support where conventional access is impeded. The versatility of hybrid drones makes them valuable assets in military operations, where they can perform intelligence, surveillance, and reconnaissance (ISR) missions with the flexibility to adapt to changing mission parameters (Vohra et al., 2023).

The classification of drones by size, usage, and design highlights the diversity and versatility of these unmanned aerial vehicles. Each type of drone has unique features and applications that make it suitable for specific tasks, contributing to their growing importance in various industries. The continuous advancement of drone technology, including improvements in propulsion systems, battery life, sensor integration, and autonomous capabilities, is expanding the potential uses and effectiveness of drones across multiple sectors. As regulations evolve and public acceptance increases, drones are poised to play an even more significant role in areas such as agriculture, infrastructure development, environmental conservation, and emergency services.

Understanding the key differences between drone types is not only essential for operators and industry professionals but also for policymakers and stakeholders who are shaping the future landscape of unmanned aviation. By appreciating the strengths and limitations of each drone category, better decisions can be made regarding deployment strategies, investment in research and development, and the establishment of regulations that promote safety while fostering innovation.

1.3 DRONE TECHNOLOGY AND COMPONENTS

1.3.1 Flight Control Systems

Flight control systems are the backbone of drone operations, enabling the aerial vehicles to maintain stability, navigate complex environments, and perform intricate maneuvers. These systems are intricate assemblies that integrate various components such as autopilot systems, remote control interfaces, and an array of sophisticated sensors. The seamless operation of these components ensures that drones can execute tasks ranging from simple flight to advanced autonomous missions.

1.3.1.1 Fundamentals of Flight Dynamics

Understanding flight control systems requires a solid grasp of flight dynamics—the study of forces and moments acting on an aircraft and how they influence its motion. Flight dynamics is crucial for designing control systems that can effectively manage the drone's behavior in response to pilot inputs and environmental disturbances.

A drone has six degrees of freedom, which include three rotational movements (rolling around the longitudinal axis, pitching around the lateral axis, and yawing around the vertical axis) and three translational movements (surging forward/backward, swaying left/right, and heaving up/down). The Newton-Euler equations, which explain the translational and rotational dynamics of rigid bodies, control these motions.

These equations form the foundation for developing mathematical models of drone behavior, which are essential for designing effective control systems. By understanding how forces and moments affect the drone's motion, engineers can create control algorithms that adjust motor speeds and control surfaces to achieve desired flight characteristics.

The translational motion of a drone can be expressed as:

$$m\ddot{\mathbf{X}} = \mathbf{F}_{\text{thrust}} + \mathbf{F}_{\text{gravity}} + \mathbf{F}_{\text{aero}} \tag{1}$$

where:

- m is the mass of the drone
- $\ddot{\mathbf{X}}$ is the acceleration vector
- $\mathbf{F}_{\text{thrust}}$, $\mathbf{F}_{\text{gravity}}$, and $\mathbf{F}_{\text{aero}}$ represent the thrust force, gravitational force, and aerodynamic forces, respectively.

The rotational motion is described by:

$$\mathbf{I}\dot{\omega} = \mathbf{M}_{\text{thrust}} + \mathbf{M}_{\text{aero}} - \omega \times (\mathbf{I}\omega) \tag{2}$$

where:

- $\mathbf{I}$ is the inertia tensor
- $\dot{\omega}$ is the angular acceleration vector
- $\mathbf{M}_{\text{thrust}}$ and $\mathbf{M}_{\text{aero}}$ are the moments generated by thrust and aerodynamic forces
- ω is the angular velocity vector

FIGURE 1.3 Equations for translational motion and rotational motion

1.3.1.2 Control Theory in Flight Systems

Flight control systems utilize control theory to maintain desired flight paths and orientations, ensuring stability and responsiveness under various operating conditions. Control theory provides the mathematical framework for designing controllers that can manage the drone's dynamic behavior.

Proportional-Integral-Derivative (PID) controllers are widely used in drone control systems due to their simplicity and effectiveness. A PID controller calculates the control input *u(t)* based on the error *e(t)* between the desired state and the actual state.

While PID controllers are effective for linear systems and small perturbations, drones often exhibit highly nonlinear dynamics and may operate in environments with significant uncertainties. To enhance performance, advanced control strategies are employed.

Model Predictive Control (MPC) is one such advanced method that optimizes control actions by predicting future system behavior over a finite horizon. MPC solves an optimization problem at each time step, considering constraints and desired performance criteria. This approach allows for anticipatory control actions that improve stability and responsiveness.

Adaptive control adjusts controller parameters in real time to handle system uncertainties or changes in the drone's dynamics. By continuously estimating system parameters, adaptive controllers maintain optimal performance even as the drone's characteristics vary due to factors like payload changes or component wear.

$$u(t) = K_p e(t) + K_i \int_0^t e(\tau)d\tau + K_d \frac{de(t)}{dt} \quad (1)$$

where:

- K_p, K_i, and K_d are the proportional, integral, and derivative gains, respectively.
- The proportional term addresses the present error.
- The integral term accounts for the accumulation of past errors.
- The derivative term predicts future errors based on the current rate of change.

FIGURE 1.4 Equations for control input

Drone operation involves nonlinear dynamics that can be handled by nonlinear control techniques like sliding mode control and feedback linearization. Through input transformation and variable modification, feedback linearization converts the nonlinear system into an analogous linear system, enabling the use of linear control techniques. By guiding the system's state along a preset sliding surface in the direction of the intended state, sliding mode control offers resilience against model uncertainties and outside disruptions.

1.3.1.3 Autopilot Systems

Autopilot systems play a central role in autonomous flight for drones to execute predetermined trajectories and perform functions independently. Autopilot systems use a combination of sensors and algorithms to control the flight dynamics of the drone and provide precision and reliability in a range of different flying situations.

The primary function of autopilot systems includes stabilizing the drone, flying to given waypoints, and performing defined flight maneuvers very accurately. Autopilots today use a blend of Global Positioning System (GPS) for global positioning navigation, Inertial Measurement Units (IMUs) for the detection of orientation, and barometric altimeters for precise altitude regulation. Through the integration of data from these sensors, the autopilot system ensures the intended flight path of the drone and corrects for external disturbances in the form of wind turbulence or magnetic interference (Bristeau et al., 2011).

To keep the drone stable and responsive while flying, advanced algorithms process sensor data in real time to adjust motor speeds and control surfaces. Autopilot systems have built-in fail-safe routines that maximize the safety of operation. For example, in the case of a communications failure between the drone and the ground station, the autopilot will automatically trigger a return-to-home maneuver and send the drone back to a pre-calculated location.

Key technologies in autopilot systems include the integration of advanced sensors and robust control algorithms. The use of low-cost inertial sensors, combined with computer vision techniques and sonar, enhances the drone's ability to navigate in environments where GPS signals are weak or unavailable, such as indoors or in urban canyons. Sensor fusion algorithms integrate data from multiple sources to provide accurate and reliable estimates of the drone's position and orientation (Zhih et al., 2015).

Autopilot platforms like ArduPilot and DJI's A3 and N3 systems offer sophisticated stabilization and navigation capabilities. These systems support various flight modes, including manual control, assisted flight, and fully autonomous operation. They also provide developers with customizable parameters, enabling fine-tuning of flight characteristics to suit specific applications.

Recent advancements in autopilot technology involve the incorporation of artificial intelligence and machine learning algorithms. These enhancements allow drones to adapt to dynamic environments, recognize objects, and make decisions based on real-time data analysis. For example, machine learning algorithms can improve obstacle avoidance capabilities by enabling the drone to learn from previous encounters with obstacles and adjust its flight path accordingly.

1.3.1.4 Remote Control Systems

Remote control systems enable pilots to manually operate drones from a distance, providing a direct and responsive link between the pilot and the drone. These systems are essential for both consumer-grade drones used in recreational activities and professional applications that require precise human control.

A typical remote control system consists of a transmitter, commonly a handheld controller, and a receiver installed on the drone. The transmitter sends control signals via radio frequencies, which the receiver decodes to adjust the drone's motors and control surfaces accordingly (Ebeid et al., 2017). The communication between the transmitter and receiver must be reliable and resistant to interference to ensure safe and effective control of the drone.

Modern remote control systems offer intuitive user interfaces designed to enhance the pilot's situational awareness and control precision. These interfaces often include high-resolution touch screens that display real-time telemetry data, such as battery levels, signal strength, altitude, and speed (Ma & Cheng, 2016). Additionally, customizable buttons and controls allow pilots to access specific functions quickly, such as camera controls, flight mode switches, and emergency procedures.

Systems like the DJI Smart Controller provide high-brightness displays that remain visible even in direct sunlight, which is essential for outdoor operations. The integration of real-time video feeds enables pilots to view the drone's perspective, facilitating tasks like aerial photography, surveillance, and inspection. This visual feedback is critical for navigating complex environments and executing precise maneuvers.

Moreover, some remote control systems are designed to work with mobile devices or tablets, offering flexible options for different operational needs. Software applications associated with these controllers often provide additional functionalities, such as flight planning, automated mission execution, and post-flight data analysis.

1.3.2 Propulsion Systems

Propulsion systems are fundamental to drones, providing the necessary thrust to achieve lift and maneuver through the air. The efficiency and reliability of the propulsion system directly impact the drone's performance, flight time, and ability to carry payloads.

1.3.2.1 Electric Motors

Electric motors are the most common propulsion systems in drones, particularly for small and medium-sized unmanned aerial vehicles (UAVs). They offer a balance of efficiency, simplicity, and control that is well-suited to the demands of modern drone applications.

There are two primary types of electric motors used in drones: brushed and brushless motors. Brushed motors are simpler and less expensive but suffer from lower efficiency and shorter life spans due to mechanical friction and wear (Hassanalian & Abdelkefi, 2017). In contrast, brushless motors have no physical brushes and commutators, resulting in higher efficiency, greater power output, and reduced maintenance requirements. Brushless Direct Current (BLDC) motors are the preferred choice for most drones due to their superior performance characteristics.

The performance of electric motors in drones is characterized by their thrust-to-weight ratio, efficiency, responsiveness, and thermal management. High-quality brushless motors, when paired with precisely designed propellers, can provide rapid acceleration and precise control over the drone's movement (Renduchintala et al., 2019). Electronic Speed Controllers (ESCs) regulate the power supplied to each motor, enabling fine-grained control of motor speed and direction. This precise control is essential for maintaining stability, especially in multirotor drones where the flight dynamics depend on the differential thrust produced by multiple motors.

Advancements in motor technology have led to the development of motors with higher power densities, allowing drones to carry heavier payloads without sacrificing performance. Innovations in materials and cooling techniques have also improved motor efficiency and longevity, contributing to longer flight times and reduced operational costs.

1.3.2.2 Internal Combustion Engines

Larger drones that need more power and longer flying times than electric motors and current battery technologies can supply internal combustion engines. Drones can carry out demanding jobs and travel farther thanks to their motors.

Two-stroke and four-stroke internal combustion engines are the two main engine types found in drones. Compared to four-stroke engines, two-stroke engines are lighter, generate greater power per unit weight, use less fuel, and emit more emissions (Nouacer et al., 2020). Four-stroke engines offer smoother operation, better fuel economy, and longer service life, making them suitable for missions where endurance and reliability are paramount.

Internal combustion engines are suitable for applications requiring extended flight times and the ability to carry heavy payloads. Examples include agricultural spraying, where drones must cover large fields; surveillance missions that demand long-duration loitering; and cargo transport, where lifting capacity is critical (Elmeseiry et al., 2021). The use of liquid fuels with higher energy densities than batteries allows these drones to operate for several hours, significantly expanding their operational capabilities.

However, drones powered by internal combustion engines face challenges such as increased mechanical complexity, higher maintenance requirements, and stricter regulatory compliance due to noise and emissions. Advances in hybrid propulsion systems are addressing some of these issues by combining the benefits of internal combustion engines with electric propulsion.

1.3.2.3 Hybrid Systems

In order to capitalize on the benefits of both technologies, hybrid propulsion systems integrate internal combustion engines and electric motors. These systems aim to enhance efficiency, extend flight times, and provide greater operational flexibility.

In a typical hybrid system, an internal combustion engine drives a generator that produces electricity to power electric motors or recharge the onboard batteries (Krishnaraj et al., 2021). This configuration allows the drone to benefit from the high energy density of liquid fuels while maintaining the precise control and responsiveness of electric motors. Hybrid systems can switch between power sources or use them simultaneously, optimizing performance based on the mission requirements.

Hybrid propulsion systems offer extended flight durations and the ability to carry heavier payloads, making them ideal for long-endurance missions. They also reduce the reliance on large battery packs, decreasing the overall weight and potentially lowering costs (Hassanalian & Abdelkefi, 2017). However, the complexity of integrating two propulsion systems presents engineering challenges. These include managing the additional weight, ensuring reliable operation of both systems, and developing sophisticated control algorithms to coordinate power distribution.

Moreover, maintenance and operational costs may increase due to the need to service both mechanical and electrical components. Careful design and thorough testing are essential to realize the benefits of hybrid systems while mitigating their drawbacks.

1.3.3 Power Sources

The power source is a crucial component of drone technology, determining the flight duration, performance capabilities, and operational efficiency. Advances in power source technologies directly influence the evolution of drone capabilities, enabling longer flight times, heavier payloads, and more energy-intensive applications.

1.3.3.1 Battery Technologies

Lithium-ion batteries are the most common power source for drones due to their high energy density, efficiency, and rechargeability. They provide a favorable balance between weight and energy capacity, making them suitable for a wide range of drone applications (Dutta & Goswami, 2020). The performance of lithium-ion batteries is characterized by their voltage stability, charge-discharge efficiency, and lifecycle longevity, which are essential for reliable drone operations.

Energy density and charging speeds have improved due to advancements in chemistry and designs. Silicon anodes and solid-state electrolytes are two technologies that have the potential to enhance capacity for energy storage along with safety. But to avoid hazards caused by overcharge, heat, and thermal runaway, lithium-ion batteries need to be managed. Battery Management Systems (BMS) form the core of monitoring the battery's condition, balancing the cell voltage levels, and guaranteeing secure use through the whole life cycle of the drone.

Alternative battery technologies, such as lithium-polymer (Li-Po) and solid-state batteries, are being explored to overcome the limitations of traditional lithium-ion batteries (Subramanian et al., 2021). Li-Po batteries offer higher power output and can be manufactured in various shapes and sizes, providing design flexibility for different drone configurations. Solid-state batteries promise greater safety due to the elimination of liquid electrolytes, reducing the risk of leakage and combustion. Researchers are also investigating lithium-sulfur and lithium-air batteries, which have the potential to significantly increase energy density, although challenges related to stability, lifespan, and manufacturing scalability remain.

1.3.3.2 Fuel Cells

Fuel cells are an emerging power source for drones, offering longer flight times and higher energy efficiency compared to conventional batteries. The most popular kind of fuel cells found in drones are hydrogen fuel cells, which use a chemical interaction between hydrogen and oxygen to produce energy with water as the only byproduct (Elmeseiry et al., 2021). Fuel cells provide high energy density and can continuously generate electricity as long as fuel is supplied, enabling significantly longer flight durations compared to battery-powered systems.

Fuel cells are particularly useful in applications requiring long-endurance flights, such as surveying large geographic areas, environmental monitoring, and extended surveillance missions (Nouacer et al., 2020). The quiet operation and low emissions of fuel cell-powered drones make them suitable for sensitive environments and operations where minimal disturbance is required. However, challenges associated with fuel cell technology include the storage and handling of hydrogen fuel, the cost of fuel cell systems, and the need for infrastructure to support refueling operations. Advances in fuel storage methods, such as high-pressure tanks and metal hydride storage, and reductions in system costs are necessary for wider adoption in the drone industry.

1.3.3.3 Solar Power

Solar power is being explored as a sustainable power source for drones, especially for high-altitude, long-endurance missions where sunlight availability is consistent. Solar panels can be integrated into the drone's wings or body to harness sunlight and convert it into electrical energy (Ridwan & Alfindo, 2019). High-efficiency photovoltaic cells are required to generate sufficient power for the drone's propulsion and onboard systems. The integration of solar panels must be carefully designed to minimize additional weight and aerodynamic drag, which can affect flight performance.

Solar-powered drones have the potential for extended or even perpetual flight times during daylight hours, making them ideal for applications such as atmospheric research, telecommunications relay platforms, and environmental monitoring (Tkáč & Mésároš, 2019). However, the efficiency of current solar cells limits the power available, affecting the drone's payload capacity and flight speed. Energy storage systems, such as batteries or supercapacitors, are necessary to maintain operations during periods without sunlight, such as nighttime or cloudy conditions. Advancements in solar cell efficiency and lightweight energy storage solutions are critical to overcoming these limitations and realizing the full potential of solar-powered drones.

1.3.3.4 Nuclear Power Sources

An emerging area in drone power technologies is the development of nuclear power sources, specifically nuclear batteries, which promise exceptionally long endurance and high energy density. A Chinese startup has been developing nuclear-powered batteries for drones, utilizing radioisotope thermoelectric generators (RTGs) that convert heat released by the decay of radioactive isotopes into electricity. These nuclear batteries can potentially provide continuous power for years without the need for refueling, significantly extending the operational capabilities of drones.

The use of nuclear power sources in drones could revolutionize applications that require ultra-long endurance, such as deep-space exploration, remote sensing in inaccessible areas, and continuous environmental monitoring. The high energy density of nuclear batteries allows drones to carry heavier payloads or operate energy-intensive equipment without compromising flight duration.

However, the implementation of nuclear power sources in drones presents significant challenges and concerns. Safety is paramount, as the handling and potential release of radioactive materials pose risks to humans and the environment. Strict regulatory compliance is required to ensure that the design, manufacturing, and operation of nuclear-powered drones meet international safety standards.

To overcome these obstacles, developments in nuclear engineering and materials science are crucial. The development of safe containment methods, efficient thermoelectric materials, and robust shielding can mitigate risks associated with radiation exposure. Research into alternative isotopes with lower radiation profiles and longer half-lives may also enhance the viability of nuclear batteries for drone applications.

The exploration of various power sources for drones reflects the ongoing quest to enhance flight duration, performance, and operational efficiency. Battery technologies continue to evolve, with innovations aiming to increase energy density and safety. Fuel cells offer promising alternatives for long-endurance missions, while solar power provides sustainable energy solutions for specific applications. The potential introduction of nuclear power sources represents a significant leap in drone capabilities, though it brings complex challenges that must be carefully managed.

Choosing the right technology for a given drone application requires an understanding of the advantages and disadvantages of each power source. The future of drone operations will be influenced by the developments in this field as research and development continue, opening up new avenues and broadening the scope of unmanned aerial systems.

1.3.4 Navigation and GPS Systems

Navigation and GPS systems are essential for ensuring that drones can accurately and safely reach their intended destinations, execute complex flight patterns, and operate autonomously in various environments.

1.3.4.1 Global Positioning System (GPS)

GPS is a widely used technology for drone navigation, providing precise location and timing information critical for flight control and mission execution.

GPS operates by receiving signals from a constellation of satellites orbiting the Earth. By calculating the time it takes for signals from multiple satellites to reach the receiver, the drone's GPS module can determine its exact position, velocity, and altitude (Sander et al., 2018). This information is integral to navigation, enabling the drone to follow predetermined flight paths, hover at specific locations, and return to its launch point autonomously.

Factors including air conditions, multipath errors—where signals bounce off surfaces before reaching the receiver—and signal interference from buildings or topography can all have an impact on GPS accuracy. Advanced GPS systems employ augmentation techniques such as Real-Time Kinematic (RTK) positioning and Differential GPS (DGPS) to improve accuracy (Ebeid et al., 2017). These techniques are appropriate for precise applications including surveying, agriculture, and infrastructure inspection because they offer centimeter-level accuracy.

Integrating GPS data with other sensor inputs, such as IMUs and magnetometers, further improves navigation reliability and robustness.

1.3.4.2 Inertial Navigation Systems (INS)

INS are used in conjunction with GPS to provide accurate navigation data, particularly in environments where GPS signals are unreliable or unavailable.

Accelerometers and gyroscopes make up an INS, which measures the drone's angular velocity and acceleration along several axes (Nouacer et al., 2020). The system determines the drone's relative position and orientation over time by analyzing this data. When GPS signals are lost or deteriorated, the drone can still navigate accurately thanks to a technique called dead reckoning.

By integrating INS with GPS, drones achieve highly accurate and reliable navigation. The INS provides continuous position updates with high bandwidth and low latency, while GPS offers absolute positioning to correct any drift that accumulates in the INS data over time (Sander et al., 2018). This synergistic relationship enhances the drone's ability to navigate complex environments, maintain stability, and execute precise maneuvers.

Advanced navigation algorithms, such as Kalman filters, are employed to fuse data from GPS, INS, and other sensors, optimizing the accuracy and reliability of the navigation solution.

1.3.4.3 Collision Avoidance Systems

Collision avoidance systems are critical for ensuring the safety of drones during flight, particularly in environments with obstacles, other aircraft, or dynamic hazards.

A range of sensors are used by collision avoidance systems to identify obstructions in the drone's path. In order to calculate the distance to surrounding objects, ultrasonic sensors send out sound waves and time how long it takes for echoes to return (Zhang & Chandramouli, 2019). Infrared sensors detect obstacles by measuring reflected infrared light. Light Detection and Ranging (LIDAR) systems use laser pulses to create detailed 3D maps of the environment. Additionally, stereo vision cameras and radar systems provide depth perception and object detection capabilities.

Advanced algorithms process sensor data in real time to identify potential collisions and compute avoidance maneuvers (Ma & Cheng, 2016). These algorithms may employ techniques from robotics and artificial intelligence, such as simultaneous localization and mapping (SLAM), path planning, and machine learning. By predicting the trajectories of moving obstacles and the drone itself, the system can adjust the flight path proactively to maintain safe distances.

Implementing collision avoidance is particularly challenging in dynamic environments or when operating at high speeds. Regulatory frameworks are evolving to mandate the inclusion of such systems in commercial drones, emphasizing their importance in ensuring airspace safety.

1.3.5 Sensors and Cameras

Sensors and cameras are essential components of drones, enabling them to capture data, perceive their environment, and perform a wide array of tasks across different industries. The integration of advanced sensor technologies enhances the capabilities of drones, allowing for improved situational awareness, precise measurements, and sophisticated data acquisition. These systems are critical for applications ranging from aerial imaging and environmental monitoring to industrial inspection and autonomous navigation.

1.3.5.1 Visual Cameras

Visual cameras are widely used in drones for applications such as aerial photography, videography, surveillance, and inspection. They serve as the primary means for capturing detailed imagery and video content from aerial perspectives, providing valuable information utilized in various sectors.

Drones can be equipped with different types of cameras, including fixed-mount cameras and gimbal-stabilized cameras (Subramanian et al., 2021). Fixed-mount cameras are directly attached to the drone's frame and are suitable for applications where the camera's field of view remains constant. However, they are limited in their ability to compensate for the drone's movements, which can result in less stable imagery. Gimbal-stabilized cameras are mounted on motorized platforms that actively compensate for the drone's pitch, roll, and yaw movements, providing smooth and stable imagery even during dynamic flight. This stabilization is crucial for professional-grade aerial imaging, where image clarity and stability are paramount.

High-resolution sensors and optical zoom capabilities enhance the quality and versatility of the captured images and videos. Modern drone cameras often feature sensors with resolutions ranging from 12 to over 100 megapixels, enabling detailed imagery suitable for cinematic productions and detailed inspections. Optical zoom allows the camera to focus on distant subjects without loss of image quality, which is essential for applications like surveillance or infrastructure assessment.

Drones with visual cameras are able to accomplish activities including filming high-definition aerial footage for media creation, offering previously unattainable or costly views (Yu, 2023). Without the need for labor-intensive or potentially dangerous physical inspections, drones fitted with high-resolution cameras can closely check infrastructure, such as electricity lines and bridges, to find flaws or places that need maintenance.

The data collected by visual cameras can be processed using computer vision algorithms to extract valuable information and support decision-making processes. Techniques such as image segmentation, object detection, and pattern recognition enable automated analysis of large datasets, reducing the time and effort required for manual interpretation.

1.3.5.2 Infrared Cameras

Infrared cameras detect thermal radiation and are used in applications where temperature differences are of interest, such as search and rescue, firefighting, and industrial inspections. By capturing the infrared energy emitted by objects, these cameras provide thermal images that reveal temperature variations, enabling the detection of anomalies and enhancing situational awareness.

Infrared cameras capture images based on the heat emitted by objects, allowing visualization of temperature variations that are not visible to the human eye (Zhang Chandramouli, 2019). The thermal radiation is converted into electronic signals that generate images representing temperature distribution across the observed scene. Warmer objects appear brighter, while cooler objects appear darker, facilitating the identification of temperature-related phenomena.

In search and rescue operations, drones equipped with infrared cameras can locate missing persons by detecting body heat, even in low-visibility conditions. This capability significantly enhances the effectiveness and efficiency of rescue missions, reducing response times and increasing the chances of survival for individuals in distress.

Industrial inspections benefit from the ability to detect thermal irregularities in electrical installations, pipelines, and mechanical systems, enabling preventative maintenance and reducing the risk of failures (Subramanian et al., 2021). Overheating components can be early indicators of malfunctions or impending failures. By identifying these issues through thermal imaging, maintenance can be scheduled proactively, avoiding costly downtime and enhancing safety.

1.3.5.3 LIDAR and RADAR Systems

Light Detection and Ranging (LIDAR) and Radio Detection and Ranging (RADAR) systems are advanced sensing technologies that apply to mapping, the detection of obstacles, and collision avoidance. They enhance the extent to which the drone can sense the world in three dimensions to provide accurate measurements and real-time situational awareness.

LIDAR uses laser pulses to measure distances to objects, creating high-resolution 3D representations of the environment (Nouacer et al., 2020). The LIDAR sensor emits rapid pulses of laser light that reflect off surfaces and return to the sensor. By measuring the time it takes for the light to return, the system calculates the distance to each point, generating a point cloud that represents the spatial characteristics of the environment.

This technology is essential for applications requiring precise spatial data, such as topographic mapping and autonomous navigation. In topographic mapping, LIDAR-equipped drones can survey large areas quickly, producing detailed elevation models that are invaluable for construction planning and geological studies. In autonomous navigation, LIDAR provides real-time data for obstacle detection and path planning, enabling drones to navigate complex environments safely.

RADAR systems emit radio waves that reflect off objects, providing information about their distance, speed, and movement. RADAR is effective in various weather conditions and can detect objects at longer ranges compared to optical systems. Unlike LIDAR, which may be affected by atmospheric conditions such as fog or dust, RADAR can penetrate through these obstructions, making it suitable for all-weather operations.

RADAR systems enhance situational awareness by detecting obstacles and other aircraft, contributing to safe operation in congested airspace or adverse weather conditions (Zhang & Chandramouli, 2019). They are particularly valuable in applications where safety and reliability are paramount.

Integrating LIDAR and RADAR sensors with advanced processing algorithms allows drones to operate more autonomously and perform complex tasks with higher levels of safety and efficiency. Sensor fusion techniques combine data from multiple sources to improve accuracy and robustness. The use of these advanced sensing technologies in drones supports applications such as infrastructure inspection, where detailed 3D models of structures are required for analysis and maintenance planning.

Sensors and cameras are integral to the functionality and versatility of drones, enabling them to perform a multitude of tasks with precision and efficiency. The advancements in visual cameras, infrared imaging, and LIDAR and RADAR systems have significantly expanded the capabilities of drones, allowing them to capture detailed data, navigate complex environments, and operate autonomously in diverse conditions.

The integration of these technologies facilitates applications ranging from media production and industrial inspection to emergency response and environmental monitoring. By leveraging advanced sensors, drones can provide valuable insights, enhance operational safety, and contribute to more informed decision-making processes.

As the technology continues to evolve, ongoing research and development aim to further miniaturize sensors, improve their performance, and reduce costs. The future of drones will be shaped by the continued integration of sophisticated sensors and cameras, driving innovation and expanding the horizons of what can be achieved through unmanned aerial systems.

1.3.6 Communication Systems

Communication systems are fundamental to drone operations, enabling the transmission of control commands, telemetry data, and payload information between the drone and ground control stations or other networks. Effective communication ensures that drones can be operated safely, efficiently, and in compliance with regulatory requirements. The design of these systems must consider factors such as range, bandwidth, latency, security, and reliability.

1.3.6.1 Radio Frequency (RF) Communication

Radio Frequency (RF) communication is the most prevalent method for controlling drones and transmitting data over short to medium ranges. RF systems are essential for real-time control inputs, telemetry data transmission, and high-definition video streaming from the drone to the operator.

Drones utilize various RF bands for communication, including the 2.4 GHz and 5.8 GHz Industrial, Scientific, and Medical (ISM) bands (Jun et al., 2019). These frequencies offer a balance between range, data transmission speed, and antenna size. Some systems operate in the 900 MHz band to achieve longer ranges at the expense of data throughput. The choice of frequency band affects the propagation characteristics, penetration capabilities, and susceptibility to interference.

The performance of an RF communication link can be analyzed using the Friis transmission equation, which relates the received power Pr to the transmitted power Pt, the gains of the transmitting and receiving antennas Gt and Gr, the wavelength *lambda*, and the distance d between the transmitter and receiver

This equation assumes free space propagation and does not account for obstacles, multipath effects, or atmospheric absorption. In practical scenarios, the link budget must include additional factors such as path loss exponent n, shadowing, and fading. The received power can be expressed in logarithmic form to facilitate link budget calculations.

Advanced communication protocols, such as Frequency Hopping Spread Spectrum (FHSS) and Direct Sequence Spread Spectrum (DSSS), enhance signal robustness and resistance to interference by spreading the signal over a wider bandwidth (Ebeid et al., 2017). These techniques reduce the impact of narrowband interference and make the communication link more secure.

$$P_r = P_t G_t G_r \left(\frac{\lambda}{4\pi d}\right)^2$$

FIGURE 1.5 Friis transmission equations

$$P_r \text{ (dBm)} = P_t \text{ (dBm)} + G_t \text{ (dBi)} + G_r \text{ (dBi)} - 10n\log_{10}(d) + C$$

where:

- C represents other losses or gains in the system.

FIGURE 1.6 Equation for free space propagation

The capacity of an RF communication channel is governed by the Shannon-Hartley theorem, which defines the maximum data rate *C* achievable over a communication channel with bandwidth *B* and signal-to-noise ratio *SNR*.

This relationship highlights the trade-off between bandwidth, data rate, and signal quality. To support high-definition video streaming and large telemetry data, drones require communication systems that offer high bandwidth and reliable signal integrity.

Encryption and authentication mechanisms are employed to secure the communication link against unauthorized access and potential cyber threats. Techniques such as Advanced Encryption Standard (AES) and secure key exchange protocols ensure that sensitive information remains confidential and that control commands are not tampered with.

Latency is another critical parameter, especially for control signals that require real-time responsiveness. Low-latency communication is essential for safe operation, particularly in environments with potential obstacles or other aircraft.

1.3.6.2 Cellular Communication

Recent advancements have seen the integration of cellular networks, such as 4G LTE and 5G, into drone communication systems. Cellular communication offers wide-area coverage, high data rates, and low latency, making it suitable for beyond-visual-line-of-sight (BVLOS) operations.

Using cellular networks, drones can leverage existing infrastructure to maintain connectivity over long distances without the need for specialized equipment. The 5G network, with its enhanced Mobile Broadband (eMBB), Ultra-Reliable Low-Latency Communications (URLLC), and massive Machine-Type Communications (mMTC), provides the capabilities needed for advanced drone operations.

However, challenges include network availability, regulatory compliance, and potential interference with terrestrial users. Network slicing and quality of service

$$C = B \log_2(1 + \text{SNR})$$

FIGURE 1.7 Equation based on Shannon-Hartley theorem

(QoS) management are employed to prioritize drone communication traffic and ensure reliable performance.

1.3.6.3 Satellite Communication

Satellite communication is employed for long-range and beyond-visual-line-of-sight (BVLOS) operations, providing global coverage and reliable connectivity, especially in remote areas where terrestrial communication infrastructure is lacking.

Satellite communication involves transmitting data between the drone and a satellite in orbit, which then relays the data to a ground station or another satellite (Sekander et al., 2017). This method enables drones to operate over vast distances and in challenging environments, such as maritime or mountainous regions.

The propagation delay in satellite communication is significant due to the large distances involved, especially with geostationary satellites located approximately 35,786 kilometers above the Earth's surface. The latency T can be approximated by:

where h is the altitude of the satellite and c is the speed of light in vacuum (approximately 3 x 10^8 meters per second). For geostationary satellites, the round-trip latency is about 240 milliseconds, which can impact real-time control operations.

Low Earth Orbit (LEO) satellite constellations, such as SpaceX's Starlink and OneWeb, orbit at altitudes between 500 and 2,000 kilometers, significantly reducing latency to around 30–50 milliseconds. These systems offer higher data rates and lower latency, making them more suitable for drone communication.

Challenges associated with satellite communication include the cost of satellite bandwidth, the need for specialized and often heavier equipment on the drone, and regulatory considerations for frequency allocation and transmission power.

1.3.6.4 Hybrid Communication Systems

To optimize performance and cost-effectiveness, hybrid communication systems combine multiple communication technologies, such as RF, cellular, and satellite links. These systems can dynamically select the best communication link based on factors like signal quality, bandwidth requirements, and operational context.

For example, a drone might use RF communication for short-range operations, switch to cellular networks when operating within cellular coverage areas, and rely on satellite communication when in remote locations. Seamless handover mechanisms are critical to maintaining uninterrupted connectivity.

1.3.6.5 Regulatory and Security Considerations

Communication systems have to meet the standards defined by the Federal Communications Commission (FCC) in the United States or the European Telecommunications Standards Institute (ETSI) in the EU. Frequency assignment,

$$T = \frac{2h}{c}$$

FIGURE 1.8 Equation for latency

power limits for transmission, and the use of the spectrum are regulated to avoid interference in other services.

Security is paramount to prevent unauthorized access, data breaches, and control hijacking. In addition to encryption, techniques like intrusion detection systems and secure authentication protocols are implemented to safeguard communication links.

1.3.6.6 Future Trends in Drone Communication

The evolution of communication technologies will significantly impact drone capabilities. The deployment of 5G networks with network slicing and edge computing supports low-latency and high-bandwidth applications. Research into millimeter-wave (mmWave) frequencies and terahertz communication offers the potential for even higher data rates, albeit with challenges related to propagation and atmospheric absorption (Zhang & Chandramouli, 2019).

Unmanned Aircraft Systems Traffic Management (UTM) systems are being developed to coordinate drone operations, requiring standardized communication protocols and interoperability among different drone platforms and service providers.

Integration into Internet of Things (IoT) systems enables drones to function as part of a larger system, communicating data to other systems and devices. Such connectivity facilitates applications ranging from sensing the environment to monitoring urban infrastructure and multi-drone coordination.

1.3.6.7 Mathematical Modeling of Communication Channels

Understanding the behavior of communication channels is essential for designing reliable drone communication systems. The path loss Lp in a wireless communication channel can be modeled using the log-distance path loss model.

The signal-to-noise ratio (SNR) is an important measure of the quality of an RF communication link. It is defined by the following equation.

The Bit Error Rate (BER) is a critical parameter that quantifies the reliability of the communication link. For a binary phase-shift keying (BPSK) modulation scheme, the BER over an additive white Gaussian noise (AWGN) channel is:

These equations provide a foundation for analyzing and optimizing the communication systems used in drones, taking into account factors like modulation schemes, coding techniques, and channel conditions.

Communication systems are a vital component of drone technology, enabling the exchange of information necessary for control, navigation, and data acquisition. Advances in RF communication, cellular networks, and satellite systems have expanded the operational capabilities of drones, allowing for longer ranges, higher data rates, and more reliable connections. Mathematical models and equations

$$L_p(d) = L_0 + 10n \log_{10}\left(\frac{d}{d_0}\right) + X_\sigma$$

where:

- L_0 is the path loss at the reference distance d_0,
- n is the path loss exponent,
- d is the distance between the transmitter and receiver,
- X_σ is a Gaussian random variable representing shadowing effects.

FIGURE 1.9 Equation for path loss

underpin the design and analysis of these systems, ensuring that drones can meet the demands of increasingly complex and diverse applications.

Understanding the intricacies of communication technologies is essential for engineers and practitioners working in the field of unmanned aerial systems. As the technology continues to evolve, ongoing research and development will address current challenges and unlock new possibilities for drone communication.

1.3.7 Software and Algorithms

Software and algorithms are the intelligence behind drone operations, enabling advanced functionalities such as autonomous flight, data analysis, and system integration.

1.3.7.1 Flight Planning Software

Flight planning software is used to create and manage flight plans for autonomous and semi-autonomous drone operations, ensuring efficient mission execution and compliance with regulations.

Flight planning software allows users to define waypoints, set altitude levels, specify flight paths, and configure mission parameters such as speed and payload operations (Nouacer et al., 2020). Advanced features include terrain following, obstacle avoidance integration, and geofencing to restrict the drone's operational area. The software often provides simulation capabilities to preview missions and detect potential issues before deployment.

Platforms like DJI Ground Station Pro, Pix4D, and Mission Planner offer user-friendly interfaces and robust features for planning and executing complex drone missions (Ebeid et al., 2017). These solutions support various drone models and payloads, providing flexibility for different applications. Integration with mapping and Geographic Information System (GIS) tools enhances the ability to analyze mission data and generate actionable insights.

$$\text{SNR} = \frac{P_r}{N_0 B} \tag{1}$$

where:

- P_r: Received power, which represents the power level of the signal at the receiver.
- N_0: Noise power spectral density, which quantifies the noise power per unit of bandwidth.
- B: Bandwidth, which is the frequency range over which the signal is transmitted.

FIGURE 1.10 Equation for signal-to-noise ratio or SNR

$$\text{BER} = Q\left(\sqrt{2\gamma}\right)$$

FIGURE 1.11 Equation for bit error rate or BER

1.3.7.2 Data Analysis Algorithms

Data analysis algorithms process and interpret the huge volumes of data gathered by drones and turn raw data into actionable insights to support decision-making.

Algorithms for image processing recognize and differentiate features in aerial images to facilitate object identification, change detection, and anomaly detection (Subramanian et al., 2021) Supervised and unsupervised machine learning techniques are employed in pattern recognition and predictive analytics. Statistical algorithms process sensor data to detect trends, correlations, and anomalies.

For agriculture, data analysis algorithms monitor the health of crops through multispectral and hyperspectral imagery analysis to inform precision farming. For construction and infrastructure management, algorithms scan for defects and map structural changes in real time. Environmental monitoring also derives advantage through tracking wildlife population levels, measuring levels of deforestation, and levels of pollution (Ma & Cheng, 2016).

The performance of these algorithms relies on the data quality and the strength of the models involved. Breakthroughs in computational capacity and cloud processing have dramatically improved the capacity to process huge datasets that are produced by drones.

1.3.7.3 Machine Learning and AI in Drones

Drone makers today incorporate more machine learning and artificial intelligence technologies in order to increase autonomy, adaptability, and efficiency.

Neural networks, including convolutional neural networks (CNNs), and recurrent neural networks (RNNs), are utilized for object recognition, classification, and analysis of time-series data (Siddappaji & Akhilesh, 2019). Drones use reinforcement learning to acquire the best behavior through sensing the environment and enhance navigation and decision-making. Computer visionsystems help drones to make sense of visual information for functions such as detecting obstacles, mapping, and tracking targets.

The integration of AI in drones offers numerous benefits, such as improved autonomy, enhanced data processing capabilities, and more efficient operations (Siddappaji & Akhilesh, 2019). AI-powered drones can adapt to changing environments, make real-time decisions, and perform complex tasks without constant human supervision.

Challenges include the need for large datasets to train AI models, the computational resources required for processing, and ensuring the reliability and safety of AI-driven systems. Ethical considerations, such as privacy concerns and the potential for unintended behaviors, must also be addressed. Ongoing research focuses on developing more efficient algorithms, edge computing solutions, and robust validation methods to overcome these challenges.

The technology and components of drones encompass a wide range of systems and subsystems that work together to enable their versatile applications. From flight control systems that stabilize and navigate the drone to propulsion technologies that provide the necessary thrust, each component plays a critical role in the functionality and performance of drones. Advanced sensors and communication networks enhance the drone's ability to perceive its environment and interact with operators or other systems. Software and algorithms imbue drones with intelligence, enabling autonomous operation and sophisticated data analysis.

As drone technology continues to evolve, these components will become increasingly integrated and advanced, leading to new capabilities and applications. The continued development of power sources, sensor technologies, and AI algorithms will expand the potential of drones in various industries, including agriculture, infrastructure, environmental monitoring, and logistics. Understanding the fundamental components and technologies of drones is essential for appreciating their impact on modern society and for contributing to future innovations in this dynamic field.

1.4 APPLICATIONS AND USE CASES

1.4.1 Consumer Applications

Consumer drones are widely popular due to their accessibility, ease of use, and diverse applications, primarily focused on recreational activities and creative endeavors.

1.4.1.1 Recreational Flying

Recreational flying with drones has become a significant hobby for many enthusiasts. These drones are generally small, lightweight, and equipped with basic flight controls, making them ideal for beginners and hobbyists. Recreational drones are used for fun activities such as drone racing, where pilots compete by navigating their drones through obstacle courses at high speeds. This sport has grown in popularity, with organized competitions and leagues offering a platform for enthusiasts to showcase their skills (Yu, 2023).

The simplicity and affordability of recreational drones have made them accessible to a broad audience. Many drones come with features like automated flight modes, GPS stabilization, and collision avoidance, making them easy to operate even for novices. These drones often have flight times ranging from 10 to 30 minutes and can reach speeds of up to 60 mph, providing an exhilarating experience for users. The rise of drone communities and online platforms has also facilitated the sharing of knowledge and experiences among enthusiasts, further driving the popularity of recreational flying (Hildebrand, 2021).

1.4.1.2 Aerial Photography and Videography

The usage of consumer drones for aerial photography and filmmaking is among the most common. With their top-notch cameras, these drones can take breathtaking aerial photographs and films. They offer unique perspectives that were previously difficult or impossible to achieve without expensive equipment or risky maneuvers. Drones like the DJI Phantom and Mavic series have revolutionized the field by providing easy-to-use platforms with professional-grade camera capabilities (Garg, 2022).

Aerial photography drones come with features such as gimbal stabilization, which ensures smooth and stable footage even during fast movements or windy conditions. Advanced models offer 4K video recording, panoramic shots, and automated flight paths that allow users to focus on capturing the perfect shot. These capabilities have opened new creative possibilities for photographers and videographers, enabling them to produce breathtaking landscapes, real estate showcases, and dynamic event coverage. The integration of artificial intelligence in some drones also allows for automated subject tracking and obstacle avoidance, enhancing the ease of use and safety during flight (Chiang et al., 2019).

1.4.2 Commercial Applications

Commercial drones are utilized across various industries to improve efficiency, reduce costs, and enhance safety. Their ability to perform tasks that are dangerous or difficult for humans makes them invaluable tools in many sectors.

1.4.2.1 Agriculture and Crop Monitoring

Precision agriculture, made possible by drones, is revolutionizing conventional farming methods. Multispectral sensors on these drones allow them to evaluate field conditions, track crop health, and maximize fertilization and irrigation. By capturing detailed aerial imagery, drones help farmers identify areas that need attention, such as sections suffering from pests, diseases, or water stress (Dutta & Goswami, 2020).

Drones in agriculture can perform various tasks such as spraying pesticides and fertilizers, mapping fields, and analyzing soil health. This data-driven approach leads to better crop management, increased yields, and reduced resource waste. For instance, drones can create detailed 3D maps of fields, enabling precise planning and management of resources. They can also monitor crop growth and predict harvest times, providing valuable insights for farmers to make informed decisions (Bharti et al., 2020).

1.4.2.2 Real Estate and Construction

In the real estate and construction industries, drones provide significant benefits by offering unique aerial views of properties and construction sites. Aerial footage captured by drones can highlight property features, surrounding landscapes, and neighborhood amenities, providing a comprehensive view that traditional ground photography cannot offer. This capability is especially useful for marketing purposes, as it helps real estate agents showcase properties more effectively (Cvitanić, 2020).

Drones are also used for site inspections and progress monitoring in construction projects. They can capture high-resolution images and 3D models of construction sites, enabling project managers to track progress, identify potential issues, and ensure compliance with safety regulations. This aerial perspective allows for more accurate and efficient site assessments, reducing the need for manual inspections and minimizing project delays (Otto et al., 2018).

1.4.2.3 Environmental Monitoring

Drones are essential for conservation and environmental monitoring. They are employed to monitor deforestation, follow wildlife, and evaluate the condition of ecosystems. Researchers and environmentalists can benefit greatly from the rich data that drones with cameras and sensors can gather on plant and animal populations (Torresan et al., 2017).

For example, drones can be used to monitor the migration patterns of wildlife, assess the impact of natural disasters, and track changes in land use. This information helps in the development of strategies for conservation and environmental protection. Drones are also used in forest management to map tree densities, monitor forest health, and detect illegal logging activities. Their ability to access remote and difficult-to-reach areas makes them ideal tools for environmental monitoring (Ghamari et al., 2022).

1.4.3 Industrial Applications

Industrial drones are designed for heavy-duty tasks and can operate in harsh environments, making them valuable tools in various industrial sectors.

1.4.3.1 Infrastructure Inspection

The use of drones to monitor vital infrastructure, such as electrical lines, pipelines, and bridges, is growing. They reduce the need for human inspections and increase safety by offering high-resolution photos and data that can spot possible problems and structural damage. For example, drones can inspect hard-to-reach areas on tall structures, minimizing the risk to human inspectors and reducing downtime (Vohra et al., 2023).

Using drones for infrastructure inspection offers several advantages, including faster data collection, lower costs, and increased accuracy. Drones equipped with thermal cameras can detect heat anomalies in power lines and pipelines, indicating potential failures that require maintenance. This proactive approach helps in preventing costly repairs and ensures the integrity and safety of critical infrastructure (Elmeseiry et al., 2021).

1.4.3.2 Energy Sector

In the energy sector, drones are used for inspecting wind turbines, solar panels, and oil rigs. They help identify potential issues and perform maintenance tasks more efficiently, reducing downtime and operational costs. For example, drones equipped with thermal cameras can detect hot spots in solar panels, indicating areas that may be malfunctioning and require maintenance (Tkáč & Mésároš, 2019).

Drones are also used to inspect underwater structures in offshore oil rigs, providing valuable data without the need for divers. This capability enhances safety and reduces the risks associated with underwater inspections. In wind energy, drones are used to inspect turbine blades for damage and wear, allowing for timely maintenance and reducing the risk of catastrophic failures. Overall, the use of drones in the energy sector improves efficiency, reduces costs, and enhances safety (Ghamari et al., 2022).

1.4.4 Military and Defense Applications

Military drones are used for various defense and combat purposes, providing significant strategic advantages.

1.4.4.1 Surveillance and Reconnaissance

Military drones are widely used for intelligence gathering and battlefield surveillance. They provide real-time data and imagery, enhancing situational awareness and decision-making. Drones such as the MQ-9 Reaper can loiter over areas of interest for extended periods, capturing high-resolution images and video that are crucial for identifying enemy positions and movements (Tsach et al., 2010).

Advanced sensor-equipped drones can follow and identify targets, keep an eye on troop movements, and give ground forces real-time intelligence. Military leaders are better equipped to plan operations and make well-informed judgments because of this skill. The use of drones for surveillance and reconnaissance has transformed modern warfare by providing a persistent and reliable source of intelligence (Sullivan, 2005).

1.4.4.2 Combat and Strike Missions

Combat drones are used for targeted strikes and combat operations. They can carry various payloads, including missiles and bombs, to engage enemy targets while minimizing risks to personnel. These drones can be remotely operated from safe locations, allowing military forces to conduct precision strikes with minimal collateral damage (Franke, 2014).

The use of combat drones has transformed modern warfare by providing a platform for executing high-risk missions without exposing pilots to danger. Drones such as the MQ-1 Predator and MQ-9 Reaper have been used extensively in conflicts around the world, demonstrating their effectiveness in both surveillance and strike roles. The ability to deploy drones for combat operations has provided military forces with a flexible and responsive tool for addressing threats and achieving strategic objectives (Kindervater, 2016).

Applications for drones are numerous and span a variety of industries, from commercial and consumer to military and industrial. They are essential instruments for improving productivity, safety, and strategic capabilities because of their adaptability, effectiveness, and capacity to carry out activities in demanding conditions.

1.5 FUTURE TRENDS IN DRONE TECHNOLOGY

1.5.1 Technological Innovations

The future of drone technology is poised to be driven by several key innovations that promise to enhance the capabilities and expand the applications of UAVs. These technological advancements include improvements in battery life, advancements in artificial intelligence (AI) and autonomous systems, and the integration of next-generation communication networks such as 5G and 6G.

1.5.1.1 Advancements in Battery Life

The major limitation of drone technology is that most drones can only fly for 20 to 50 minutes at a time, which limits how far and how long they can be used. That said, a great deal of work is going into developing battery technology that will allow them to fly for longer. Solid-state or lithium-sulfur batteries and fuel cells have longer ranges and last longer, compared to what lithium-ion batteries are currently capable of (Dutta & Goswami, 2020).

Solid-state batteries, for example, can provide higher storage of energy while being safer too because they have solid electrolytes. The new battery technology of course delivers these high-output drones with longer stints of flight and more rapid charging cycles—two fundamentals that could really help improve the workflow of drones currently in operation. Proponents of hydrogen fuel cells and their ability to achieve much longer flight times is another PEM nirvana. Fuel cells provide a clean and efficient power source for drones, capable of converting hydrogen into electricity through a chemical reaction with oxygen (Chan et al., 2018).

1.5.1.2 Improvements in AI and Autonomous Systems

The development of autonomous drones is becoming more and more dependent on artificial intelligence and machine learning. Drones can now navigate, avoid obstacles, and recognize targets with little assistance from humans thanks to AI systems. These features are especially useful in applications such as search and rescue and surveillance, and agriculture, where drones can autonomously monitor large areas and make real-time decisions based on data collected by onboard sensors (Nouacer et al., 2020).

Advancements in computer vision and sensor fusion are also enhancing the ability of drones to operate autonomously in complex environments. Drones equipped with advanced cameras, LIDAR, and RADAR systems can create detailed 3D maps of their surroundings, enabling precise navigation and obstacle avoidance. AI-powered drones can also learn from their environment, improving their performance over time through machine learning algorithms. These developments are paving the way for fully autonomous drones capable of carrying out intricate missions without human oversight (Ma & Cheng, 2016).

1.5.1.3 Integration of 5G and 6G

The integration of 5G and future 6G communication networks is expected to have a profound impact on drone operations. These next-generation networks offer ultra-low latency, high bandwidth, and reliable connectivity, which are essential for real-time data transmission and control of drones. 5G networks, for example, enable drones to transmit high-resolution video feeds and sensor data to ground stations with minimal delay, facilitating applications such as real-time surveillance, remote inspections, and live broadcasting (Han et al., 2021).

6G, which is still in the conceptual phase, promises even greater enhancements in speed, capacity, and connectivity. It is expected to support massive IoT (Internet of Things) deployments, allowing for the integration of large fleets of drones that can operate collaboratively. The high data rates and low latency of 6G will enable advanced applications such as swarm intelligence, where multiple drones can work together to perform tasks more efficiently. This collaborative approach can be particularly useful in areas like disaster response, environmental monitoring, and logistics (Chang et al., 2021).

1.5.2 Regulatory and Legal Developments

As drone technology advances, regulatory frameworks must evolve to address new challenges and ensure safe and responsible use. Regulatory and legal developments are crucial for integrating drones into national airspace, protecting privacy, and ensuring security.

1.5.2.1 Evolving Drone Laws and Regulations

Global regulatory organizations are attempting to create all-encompassing frameworks to control the growing drone usage. These rules address things like pilot qualification, airspace management, operational restrictions, and safety requirements. For example, the US Federal Aviation Administration (FAA) has established regulations for commercial drone operations, such as qualification requirements for remote pilots and prohibitions on flying at night and over people (Macias et al., 2019).

One of the significant challenges in drone regulation is integrating UAVs into controlled airspace while ensuring they do not interfere with manned aircraft. Solutions such as Unmanned Traffic Management (UTM) systems are being developed to provide real-time air traffic data and ensure safe separation between drones and other airspace users. These systems use GPS data, communication networks, and AI algorithms to manage drone traffic and prevent collisions.

1.5.2.2 Impact on Commercial Drone Use

Regulatory developments are also essential for enabling commercial drone applications. Clear and consistent regulations can provide businesses with the certainty they need to invest in drone technology. For example, regulations permitting beyond-visual-line-of-sight (BVLOS) operations can unlock new opportunities for drone delivery services, infrastructure inspections, and agricultural monitoring (Green et al., 2021).

International operations of drones are reliant on the establishment of international norms as well as the harmonization of national legislation. For the sake of fostering cross-border operations of drones and ensuring harmonization of safety standards, institutions such as the European Aviation Safety Agency (EASA) and the International Civil Aviation Organization (ICAO) are working towards the establishment of similar regulatory systems.

1.5.3 Market Growth and Economic Impact

The drone industry is experiencing rapid growth, driven by technological advancements, expanding applications, and increasing investment. This growth has significant economic implications, creating new opportunities and challenges.

1.5.3.1 Current Market Trends

The international market for drones will increase considerably in the next few years. The market will expand to more than $43 billion by 2025 due to the rising use of drones in different industries including agriculture, construction, logistics, and defense (Kapustina et al., 2021).

In agriculture, drones are being used for precision farming, crop monitoring, and livestock management. In the construction industry, drones are employed for site surveys, progress monitoring, and safety inspections. The logistics sector is exploring the use of drones for last-mile delivery, warehouse management, and inventory tracking. The defense industry continues to invest in advanced military drones for surveillance, reconnaissance, and combat missions.

1.5.3.2 Future Market Projections

Current and future predictions for the drone market establish this as an industry that will grow based on new uses, uses-cases, and infusions of innovation. The report expects drones to become more capable with new opportunities such as AI, 5G, and blockchain technologies. For example, AI-enabled drones can carry out autonomous missions, and simultaneously 5G networks offer reliable and fast communication for real-time data and control transmission (Vergouw et al., 2016).

The economic impact of the drone industry goes far beyond direct sales and services. With a shift to drones however, the manufacturing of drones would lead to job creation in different areas such as drone manufacturing, software development, pilot (drone) training, and maintenance. Drone usage can also boost productivity, lower costs, and increase efficiency in different sectors of new economies with a direct positive output on economic growth.

Battery life, AI and autonomous systems, and next-gen communications: To dive deeper into the other topics, regulatory and legal (development) is paramount to facilitate a safe and responsible operation of drones, market growth, and economic impact explore wider opportunities as well as challenges in the drone industry. And, with further advances in technology, drones are set to take on an even more significant role in various industries that will have the potential to change how humans work, live, and connect with each other.

1.6 CHALLENGES AND LIMITATIONS

1.6.1 Technical Challenges

Despite the rapid advancements in drone technology, several technical challenges need to be addressed to fully harness the potential of drones. These challenges include battery life and power management, range and payload capacity, and reliable communication systems.

1.6.1.1 Battery Life and Power Management

One of the key technical challenges for drone technology involves battery life. Consumer and commercial drones use predominantly lithium-ion batteries that have relatively low flight times of about 20 to 30 minutes. This confines the scope and distance of a mission as well as the duration of a flight for applications that need long-distance coverage or operation for a long time.

Researchers are also exploring some alternatives that could enhance the performance of the battery, namely the use of lithium-sulfur batteries, solid-state batteries, and hydrogen fuel cells. Lithium-sulfur batteries have the ability to deliver higher energy densities that would greatly enhance flight durations. Solid-state batteries provide a better safety and capacity for storing energy. Hydrogen fuel cells are also a promising innovation as these have the ability to provide longer flight times through a chemical reaction involving the use of hydrogen and oxygen (Dutta & Goswami, 2020).

Power management is another critical aspect of extending drone flight times. Efficient power management systems can optimize energy consumption by dynamically adjusting power usage based on the drone's operational requirements. This includes managing the power distribution between various onboard systems, such as propulsion, sensors, and communication devices, to maximize efficiency and extend battery life (Chan et al., 2018).

1.6.1.2 Range and Payload Capacity

The range and payload capacity of drones are critical factors that determine their suitability for different applications. Drones with limited range and payload capacity may not be able to perform tasks that require extensive travel or the transport of heavy equipment.

Advancements in propulsion technology, such as more efficient electric motors and lightweight materials, can help improve the range and payload capacity of drones. Hybrid propulsion systems that combine electric motors with internal combustion engines are also being developed to provide longer flight times and greater lifting capabilities (Elmeseiry et al., 2021).

Increasing the range and payload capacity of drones also involves optimizing their aerodynamic design. Streamlined shapes and lightweight composite materials can reduce drag and improve fuel efficiency, allowing drones to fly longer distances and carry heavier loads (Nouacer et al., 2020).

1.6.1.3 Reliable Communication Systems

Effective communications systems are vital for the secure and efficient flying of drones, particularly for beyond-visual-line-of-sight flights. The drones use communications links to send data and receive instructions from the ground stations. Break in these signals may result in a loss of control and accidents.

The integration of 5G and 6G communication networks offers promising solutions to the challenge of reliable communication. These networks provide high bandwidth, low latency, and robust connectivity, enabling real-time data transmission and control. 5G networks, for instance, can support high-definition video streaming and low-latency control commands, enhancing the safety and effectiveness of drone operations (Han et al., 2021).

1.6.2 Regulatory Hurdles

The rapid proliferation of drones has outpaced the development of regulatory frameworks, leading to several regulatory hurdles that must be addressed to ensure the safe and responsible use of drones.

1.6.2.1 Airspace Regulations

For safe drone operations and to avoid drone-manned aircraft crashes, airspace restrictions are essential. Guidelines and regulations for drone operations have been set by regulatory agencies such as the European Aviation Safety Agency (EASA) in Europe and the Federal Aviation Administration (FAA) in the United States. These rules usually include no-fly zones, limitations on aircraft heights, and the need to secure special authorization for specific kinds of operations (Macias et al., 2019).

One of the significant challenges is integrating drones into controlled airspace, where they must coexist with manned aircraft. Solutions such as Unmanned Traffic Management (UTM) systems are being developed to provide real-time air traffic data and ensure safe separation between drones and other airspace users. These systems use GPS data, communication networks, and AI algorithms to manage drone traffic and prevent collisions.

1.6.2.2 Privacy and Security Concerns

The broad use of drone technology is significantly hampered by privacy and security concerns. Concerns over surveillance and data privacy are raised by the ability of drones with cameras and sensors to take comprehensive pictures and data. Regulations that safeguard people's right to privacy while permitting the lawful use of drones for a variety of purposes are required (Green et al., 2021).

Security concerns include the potential misuse of drones for illegal activities such as smuggling, espionage, and terrorism. There is a need for robust security measures to prevent unauthorized access to drone systems and to detect and mitigate malicious activities. Technologies such as geofencing, which restricts drones from entering specific areas, and blockchain-based security solutions are being explored to address these challenges (Mehta et al., 2020).

1.6.3 Safety and Ethical Considerations

The safety and ethical implications of drone technology are critical areas that require careful consideration to ensure responsible and beneficial use.

1.6.3.1 Risk of Accidents and Failures

The risk of accidents and system failures poses significant safety concerns in drone operations. Technical malfunctions, such as communication failures, battery depletion, or software bugs, can lead to uncontrolled crashes, causing property damage, injuries, or fatalities. Ensuring the reliability and robustness of drone systems through rigorous testing, redundancy, and fail-safe mechanisms is essential to mitigate these risks (Ahmed & Sheltami, 2023).

1.6.3.2 Ethical Issues in Military Use

The use of drones in military applications raises several ethical issues, particularly regarding targeted strikes and surveillance. Drones enable remote operations, reducing the risk to military personnel but also raising concerns about accountability and decision-making in combat situations. The potential for collateral damage and civilian casualties in drone strikes necessitates strict ethical guidelines and oversight to ensure compliance with international humanitarian laws (Franke, 2014).

The use of drones for intelligence collection and monitoring is likewise subject to ethical problems. It is a difficult task to strike a balance between people's civil liberties and privacy rights and national security concerns; this calls for open policies and strong oversight procedures (Kindervater, 2016).

Addressing the technical challenges, regulatory hurdles, and safety and ethical considerations associated with drone technology is crucial for its continued development and integration into various sectors. By overcoming these challenges, drones can realize their full potential and contribute positively to society.

NOTES

1. The weight classifications may vary based on regulatory frameworks in different countries. Always consult local aviation authorities for specific regulations about drone operations in your region.
2. The applications and descriptions provide a general overview and may overlap between categories depending on specific drone models and technological advancements.

REFERENCES

Ahmed, G. A. & Sheltami, T. (2023), 'A safety system for maximizing operated uavs capacity under regu-lation constraints', *IEEE Access* 11, 139069–139081.

Bharti, M., Deepshikha & Bharti, S. (2020), 'Drone technology as a tool for improving agricultural produc-tivity', *Journal of Soil and Water Conservation* 19, 446–451.

Bhatt, K., Pourmand, A. & Sikka, N. (2018), 'Targeted applications of unmanned aerial vehicles (drones) in telemedicine.', *Telemedicine Journal and E-health: The Official Journal of the American Telemedicine Association* 24(11), 833–838.

Bristeau, P., Callou, F., Vissière, D. & Petit, N. (2011), 'The navigation and control technology inside the AR.Drone micro UAV', *IFAC Proceedings* 44, 1477–1484.

Chan, K. W., Nirmal, U. & Cheaw, W. G. (2018), 'Progress on drone technology and their applications: A comprehensive review', *AIP Conference Proceedings* 2030, 020308.

Chang, Z., Guo, W., Guo, X., Chen, T., Min, G., Abualnaja, K. & Mumtaz, S. (2021), 'Blockchain-empowered drone networks: Architecture, features, and future', *IEEE Network* 35, 86–93.

Chiang, W.-C., Li, Y., Shang, J. & Urban, T. L. (2019), 'Impact of drone delivery on sustainability and cost: Realizing the UAV potential through vehicle routing optimization', *Applied Energy* 242, 1164–1175.

Cvitanić, D. (2020), 'Drone applications in transportation', 2020 5th International Conference on Smart and Sustainable Technologies (SpliTech) pp. 1–4.

Dutta, G. & Goswami, P. (2020), 'Application of drone in agriculture: A review', *International Journal of Chemical Studies* 8, 181–187.

Ebeid, E., Skriver, M. & Jin, J. (2017), 'A survey on open-source flight control platforms of unmanned aerial vehicle', 2017 Euromicro Conference on Digital System Design (DSD), pp. 396–402.

Ehrhard, T. P. (2010), Air Force UAVs: The Secret History. Mitchell Institute for Airpower Studies (Air Force Association).

Elmeseiry, N., Alshaer, N. & Ismail, T. (2021), 'A detailed survey and future directions of unmanned aerial vehicles (uavs) with potential applications', *Aerospace* 8(12), 363.

Franke, U. E. (2014), 'Drones, drone strikes, and US policy: The politics of unmanned aerial vehicles', *Parameters* 44(1), 121.

Garg, P. K. (2022), 'Characterisation of fixed-wing versus multirotors uavs/drones', *Journal of Geomatics* 16(2), 152–159.

Ghamari, M., Rangel, P., Mehrübeoglu, M., Tewolde, G. & Sherratt, R. (2022), 'Unmanned aerial vehicle communications for civil applications: A review', *IEEE Access* 10, 102492–102531.

Green, D. R., Karachok, A. R. & Gregory, B. (2021), 'The potential of drone technology in pandemics', In A. Rajabifard, D. Paez, & G. Foliente (Eds.), *COVID-19 Pandemic, Geospatial Information, and Community Resilience*. Boca Raton: CRC Press.

Han, T., Ribeiro, I. G., Magaia, N., Preto, J., Segundo, A. H. F. N., Macêdo, A. R. L. D., Muhammad, K. & Albuquerque, V. (2021), 'Emerging drone trends for blockchain-based 5g networks: Open issues and future perspectives', *IEEE Network* 35, 38–43.

Hassanalian, M. & Abdelkefi, A. (2017), 'Classifications, applications, and design challenges of drones: A review', *Progress in Aerospace Sciences* 91, 99–131.

Hildebrand, J. M. (2021), 'Seeing like a consumer drone', Aerial Play.

Jun, S., Shastri, A., Sanz-Izquierdo, B., Bird, D. & McClelland, A. (2019), 'Investigation of antennas inte-grated into disposable unmanned aerial vehicles', *IEEE Transactions on Vehicular Technology* 68, 604–612.

Kapustina, L., Izakova, N. B., Makovkina, E. & Khmelkov, M. (2021), 'The global drone market: main development trends', SHS Web of Conferences.

Kindervater, K. (2016), 'The emergence of lethal surveillance: Watching and killing in the history of drone technology', *Security Dialogue* 47, 223–238.

Krishnaraj, S., Kumar, R. S., Krishnan, A. G., Ganeshkumar, G., Mohan, M. & Nirmal, M. (2021), 'Aero-dynamic analysis of hybrid drone', IOP Conference Series: Materials Science and Engineering 1012.

Ma, L. & Cheng, L. (2016), 'Studies of AR drone on gesture control', 2016 3rd International Conference on Materials Engineering, Manufacturing Technology and Control, pp. 1864–1868.

Macias, M., Barrado, C., Pastor, E. & Royo, P. (2019), 'The future of drones and their public acceptance', 2019 IEEE/AIAA 38th Digital Avionics Systems Conference (DASC), pp. 1–8.

Mehta, P., Gupta, R. & Tanwar, S. (2020), 'Blockchain envisioned UAV networks: Challenges, solutions, and comparisons', *Computer Communications* 151, 518–538.

Mohsan, S., Khan, M. A., Noor, F., Ullah, I. & Alsharif, M. (2022), 'Towards the unmanned aerial vehicles (uavs): A comprehensive review', Drones 6(6), 147.

Nouacer, R., Hussein, M., Espinoza, H., Ouhammou, Y., Ladeira, M. & González, R. (2020), 'Towards a framework of key technologies for drones', *Microprocessors and Microsystems* 77, 103142.

Okulski, M. & Ławryńczuk, M. (2022), 'A small UAV optimized for efficient long-range and VTOL missions: An experimental tandem-wing quadplane drone', *Applied Sciences* 12(14), 7059.

Otto, A., Agatz, N. A. H., Campbell, J. F., Golden, B. & Pesch, E. (2018), 'Optimization approaches for civil applications of unmanned aerial vehicles (uavs) or aerial drones: A survey', *Networks* 72, 411–58.

Renduchintala, A., Jahan, F., Khanna, R. & Javaid, A. (2019), 'A comprehensive micro unmanned aerial vehicle (UAV/drone) forensic framework', Digital Investigation 30, 52–72.

Ridwan, I. & Alfindo (2019), 'The effect of use of solar panels on micro scale fixed-wing UAV type as a power recharging system', The 4th Biomedical Engineering's Recent Progress in Bio-Materials, Drugs Development, Health, and Medical Devices: Proceedings of the International Symposium of Biomedical Engineering (ISBE) 2019.

Sander, J., Kuwertz, A., Mühlenberg, D. & Müller, W. (2018), 'High-level data fusion component for drone classification and decision support in counter UAV', 10651.

Sekander, S., Tabassum, H. & Hossain, E. (2017), 'Multi-tier drone architecture for 5g/b5g cellular networks: Challenges, trends, and prospects', *IEEE Communications Magazine* 56, 96–103.

Siddappaji, B. & Akhilesh, K. (2019), 'Role of cyber security in drone technology', In K. Akhilesh & D. Möller, D. (Eds), *Smart Technologies*. Singapore: Springer.

Subramanian, K. S., Pazhanivelan, S., Srinivasan, G., Santhi, R. & Sathiah, N. (2021), 'Drones in insect pest management', *Frontiers in Agronomy* 3, 12.

Sullivan, J. M. (2005), 'Revolution or evolution? the rise of the uavs', Proceedings. 2005 International Symposium on Technology and Society, 2005. Weapons and Wires: Prevention and Safety in a Time of Fear. ISTAS 2005. pp. 94–101.

Tkáč, M. & Mésároš, P. (2019), 'Utilizing drone technology in the civil engineering', *Selected Scientific Papers - Journal of Civil Engineering* 14, 27–37.

Torresan, C., Berton, A., Carotenuto, F., Gennaro, S. F. D., Gioli, B., Matese, A., Miglietta, F., Vagnoli, C., Zaldei, A. & Wallace, L. (2017), 'Forestry applications of UAVs in Europe: A review', *International Journal of Remote Sensing* 38, 2427–2447.

Tsach, S., Tatievsky, A. & London, L. (2010), *Unmanned Aerial Vehicles (UAVs).* Portico: Encyclopedia of Aerospace Engineering. https://doi.org/10.1002/9780470686652.eae385

Valavanis, K. & Kontitsis, M. (2007), *A Historical Perspective on Unmanned Aerial Vehicles*, pp. 15–46. Dordrecht: Springer.

Vergouw, B., Nagel, H., Bondt, G. & Custers, B. (2016), 'Drone technology: Types, payloads, applications, frequency spectrum issues and future developments', In B. H. M. Custers (red.) *The Future of Drone Use* (pp. 21–45). Heidelberg: Springer.

Vohra, D., Garg, P. & Ghosh, S. (2023), 'Usage of uavs/ drones based on their categorisation: A review', *Journal of Aerospace Sciences and Technologies* 74, 90–101.

Yu, L. (2023), 'The prospect of drone applications', Applied and Computational Engineering 3(1), 300–304.

Zhang, X. & Chandramouli, K. (2019), 'Critical infrastructure security against drone attacks using visual analytics', In D. Tzovaras, et al. (eds), *Computer Vision Systems* (pp 713–722). Cham: Springer International Publishing.

Zhih, C. C., Ragavan, S. V. & Shanmugavel, M. (2015), 'Development of a simple, low-cost autopilot system for multi-rotor UAVs', 2015 IEEE Recent Advances in Intelligent Computational Systems (RAICS), pp. 285– 289.

2 Precision Agriculture Takes Flight
Drone Technology in Crop Management

Srilakshmi Prabhu and Dhanya Y Bharadwaj

2.1 INTRODUCTION

By 2050, the global population is expected to surge by 70%, reaching an estimated nine billion people. This significant increase, as projected by the Food and Agriculture Organization (FAO) and the International Telecommunication Union (ITU) (Radoglou-Grammatikis et al., 2020), poses a substantial challenge for agriculture worldwide to meet the growing food demand. Current yield improvement trends are not sufficient, necessitating either more agricultural land or enhanced productivity. India, with its population projected to hit 1.66 billion by 2050, faces similar challenges, compounded by labor shortages, limited arable land, and increasing irrigation water demand (Sankhua, R. N., 2022). The COVID-19 pandemic has worsened these issues, affecting food supply and demand and threatening food security. Therefore, farmers must explore ways to boost crop production, such as expanding agricultural land or leveraging technological advancements and better resource management.

Despite its heavy reliance on agriculture, India has been slow to adopt precision agriculture (PA) technologies for managing farm inputs. PA involves the precise application of inputs to achieve higher yields than traditional methods. It aims to enhance production efficiency, product quality, and environmental sustainability by integrating information technology and management principles. Over the past few decades, India has incorporated new technologies into PA, promising significant yield increases while minimizing external inputs—especially beneficial for small farmers in developing countries.

The essence of PA lies in using information and communication technologies to process data from various sources, enabling more effective crop management. This integration has led to "Agriculture 4.0," marking a significant transformation in farming (Javaid et al., 2022). Technologies such as Unmanned Aerial Vehicles (UAVs), remote sensing, Internet of Things (IoT), Artificial Intelligence (AI), Machine Learning (ML), and Big Data Analytics (BDA) have immense potential to revolutionize agricultural practices.

DOI: 10.1201/9781003481584-2

Drones, in particular, have gained popularity in precision agriculture, remote sensing, and photogrammetry in developed countries. They are increasingly being used in India for agricultural monitoring and decision-making. Drones allow for the monitoring of various parameters, such as environmental conditions, soil nutrient status, plant growth, irrigation management, and pest control. This technology helps farmers optimize crop yields, reduce costs, and maximize resource utilization.

The utilization of drones in agriculture has increased dramatically, revolutionizing conventional farming practices with unprecedented precision and efficiency. As modern agricultural practices continue to evolve, drones have emerged as a pivotal tool for achieving increased productivity, sustainability, and resource optimization.

This chapter looks into the numerous roles by which drones make good companions to fields such as agriculture, focusing on their applications, benefits, challenges, and future prospects. By exploring the intersection of artificial intelligence (AI), cybersecurity, and data science within the context of drone technology, this chapter aims to provide valuable insights into real-life applications and case studies that highlight the transformative potential of drones in modern agriculture.

2.2 EVOLUTION OF DRONES IN AGRICULTURE

Drones comprise a wide range of flying robots, ranging from large unmanned aerial vehicles (UAVs) capable of flying long distances to mini drones designed for operation in confined spaces (Villa et al., 2016; Kangunde et al., 2021). Other terms commonly used in different regions and contexts include Remotely Piloted Aircraft (RPA), Unmanned Aerial Vehicle (UAV), and Unmanned Aircraft System (UAS). These terms are often used interchangeably to refer to various types of unmanned flying vehicles. Drones have a rich history intertwined with agricultural pursuits owing to their high mobility, enhanced stability, low cost, and high endurance in multiple tasks. Initially developed for military reconnaissance purposes, drones found their way into civilian applications, including agriculture, as early as the 1980s (Puri et al., 2017; Otto et al., 2018; Shakhatreh et al., 2019; Rejeb et al., 2022). These early agricultural drones were rudimentary in design and functionality, primarily utilized for basic aerial imaging tasks.

Over the decades, rapid advancements in drone technology have propelled their capabilities to unprecedented heights, revolutionizing their role in agriculture. Modern agricultural drones boast cutting-edge features such as high-resolution cameras, multispectral and hyperspectral sensors, Light Detection and Ranging (LiDAR) technology, Global Positioning System (GPS) navigation systems, and autonomous flight capabilities. These technological advancements have significantly enhanced the precision, efficiency, and versatility of drones in agricultural applications.

2.3 DRONES USED IN AGRICULTURE

Drones have become increasingly relevant in agriculture due to their ability to provide valuable data and insights for farmers. They can be used for various tasks such as farm analysis, soil and field analysis, irrigation management, and crop monitoring. Drones generate 3D maps for soil analysis, which helps farmers make informed decisions during seed plowing and manage nitrogen levels for better crop growth. Drones can cover large areas of farmland efficiently, providing farmers with real-time information to optimize crop yield and prevent damage to fields. Overall, the use of drones in agriculture has the potential to revolutionize farming practices by enabling precision agriculture and improving overall productivity. The practical applications for drones in agriculture have expanded significantly, and the global drone market has continued to grow, with some estimates valuing it in the hundreds of billions of dollars by 2025 (Puri et al., 2017).

Common types of agricultural drones include fixed-wing drones, rotary-wing drones (such as quadcopters and octocopters), and hybrid drones.

(a) Fixed-Wing Drones: These drones resemble traditional airplanes and are well-suited for large-scale aerial surveys and mapping tasks. They offer extended flight times and coverage areas, making them ideal for crop monitoring and assessment over vast agricultural landscapes.
(b) Rotary-Wing Drones: Rotary-wing drones, characterized by their vertical take-off and landing capabilities, are highly maneuverable and versatile (Figure 2.1b). They excel in close-range inspections, precision spraying, and monitoring tasks, making them indispensable tools for crop management and maintenance.
(c) Hybrid Drones: Hybrid drones combine the advantages of both fixed-wing and rotary-wing designs, offering versatility in flight capabilities (Figure 2.1c). They can seamlessly transition between vertical and horizontal flight modes, making them suitable for a wide range of agricultural applications, including crop surveillance, mapping, and spraying.

Each type of drone is equipped with specific functionalities tailored to the unique requirements of agricultural operations, ranging from aerial imaging and mapping

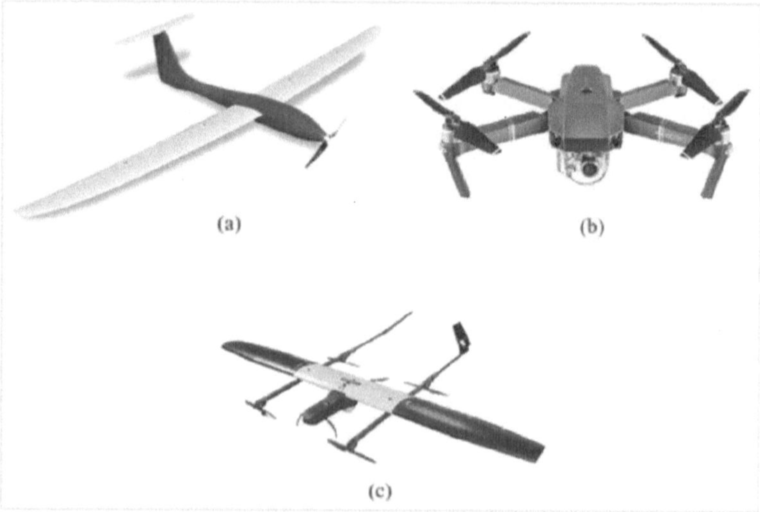

FIGURE 2.1 Different types of UAVs: (a) fixed-wing, (b) rotary wing, and (c) hybrid

to precision spraying, crop health monitoring, and beyond. These drones serve as invaluable assets in modern agriculture, empowering farmers with actionable insights and enhanced decision-making capabilities.

The capabilities of an agricultural UAS rely on that of the onboard sensors, computing unit, and other equipment. Due to the size and weight constraints associated with the UAS's payload, and air time due to limited battery capacity, drones must be chosen carefully to suit the intended mission. The key agricultural payload technologies that are available for drones are provided in Table 2.1.

Among the drone payloads mentioned in Table 2.1, the sprayer plays a vital role in agriculture as it is useful in both seed and pesticide spraying, both of which determine crop yield. The sprayer functions by breaking down the sprayed liquid, which can be a suspension, emulsion, or solution, into minuscule droplets and dispersing it with minimal force to distribute it evenly (Talaviya et al., 2020). It is also responsible for regulating the amount of pesticide to prevent over-application. Excessive pesticide use can be inefficient or detrimental to both the soil and the crop. Sprayers are categorized into four types based on the energy required to atomize and disperse the spray liquid: hydraulic energy sprayers, gaseous energy sprayers, centrifugal energy sprayers, and kinetic energy sprayers. Accordingly, agricultural drones are specifically classified into different types as shown in Figure 2.2.

TABLE 2.1
Summary of Payload Technologies Available for Agricultural Drones

RGB Camera	**Effective for creating 3D crop models, identifying individual plants, estimating biomass, and spotting diseases and weeds.**
Multispectral Camera	Captures various wavelength bands from electromagnetic spectrum: RGB—One or several bands of visual light (Red/Green/Blue). Near Infrared (NIR)—To determine physical and chemical properties of organic substances and also allows capture of Normalized Difference Vegetation Index (NDVI). Red-edge—Region of NIR with rapid change in reflectance of vegetation and is well suited to detect chlorophyll content.
Hyperspectral Camera	Collects hundreds of electromagnetic bands. Compared to multispectral cameras, the bands are narrower, and the sensor is more expensive and produces a lot more data. However, the higher spectral resolution may allow, e.g., plant phenotyping.
Thermal Camera	Captures infrared radiation to provide thermal images useful for assessing seed viability, identifying diseases, monitoring water stress, and planning irrigation.
Depth Sensor	Utilizes technologies like RGB-D and LiDAR to assist in creating 3D models and monitoring altitude, with LiDAR providing higher accuracy albeit at a greater cost and weight.
Sprayers	Enables precise spraying in challenging terrain, reducing chemical drift, and improving efficiency, and some models handle both liquid and solid substances like fertilizers and seeds.
Grippers and Manipulators	Early development stage payloads with applications in tasks such as fruit harvesting, pruning, and sample collection.

FIGURE 2.2 Types of agricultural drones

2.4 Working of Agricultural Drones

The working concept of agricultural drones involves collecting data from GPS and sensor-equipped farm equipment, which is then transmitted to a ground control station (GCS) via satellite (Figure 2.3). From there, the data are sent over the internet to users for analysis and regulation of farm implements. In real-time implementation, the GCS gathers information on drones in the fleet, including geographic data, and manages drone fleet missions. The fleet consists of multiple drones that receive missions from the GCS and collaborate to complete them. Connectivity between drones and the GCS is crucial for implementing drone-assisted wireless communications and providing instructions to field implements. Drones can also serve as standalone input applicators for site-specific management.

Let us understand the step-by-step process involved in working of agricultural drones. First, the process begins with meticulous setup and pre-flight checks, including assembly and battery charging, followed by rigorous system checks. Agricultural drones typically come disassembled for transport. Assembling involves attaching propellers, securing the payload (sensors, cameras), and ensuring all components are securely fastened. Charging the drone's batteries is crucial as flight time depends on battery life. Lithium-polymer (LiPo) batteries are common for their energy density and weight. Before flight, perform a thorough system check. This includes verifying connections, sensor functionality, GPS lock, and ensuring the flight controller and software are operational. The importance of meticulous system checks, including verifying connections, sensor functionality, GPS lock, and ensuring the flight controller and software are operational, is affirmed by several studies (Brown & Johnson, 2011). Second, flight planning is critical, involving mission definition and precise flight path design using specialized software that considers factors like altitude, speed, and sensor overlap for optimal data collection. Studies emphasize the significance of defining the flight's purpose (e.g., crop monitoring) and determining

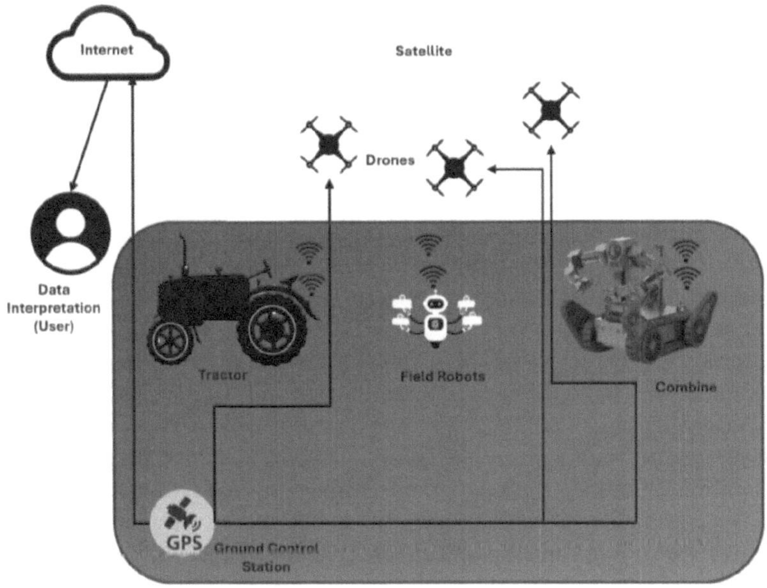

FIGURE 2.3 Working concept of UAV in agriculture

the area to be covered (Crabb et al., 2019). Third is take-off and navigation, where take-off happens either manually or through automated commands. Drones ascend to the predefined altitude and stabilize. Once airborne, agricultural drones navigate autonomously along predefined paths, utilizing GPS for positioning and sensors for altitude and orientation. The autonomous navigation feature is critical for maintaining flight paths and ensuring comprehensive data collection (Greenfield, 2022). Fourth is the crucial step of data collection, where these drones during flight employ a variety of sensors to capture data relevant to agricultural tasks, including RGB, multispectral, and thermal imagery, as well as environmental parameters like temperature and humidity. This is monitored in real time by operators to ensure quality and adjust flight parameters if needed. Live feeds assist in immediate decision-making. Real-time monitoring allows operators to adjust flight parameters to ensure data quality, a process validated by multiple sources (Garcia et al., 2020).

Data processing and analysis, being the sixth step, involves transfer of collected data to a computer for processing, upon drone landing. This includes image stitching, calibration, and conversion into usable formats. Then, the software processes images to extract actionable insights. This may involve vegetation indices (Normalized Difference Vegetation Index, NDVI), plant health indicators, and anomaly detection (e.g., pest infestations). Subsequently, the analyzed data provides insights into crop health, growth patterns, and potential issues. Operators interpret findings to make informed decisions regarding irrigation, fertilization, and pest management. And the generated reports summarize findings and recommendations, which facilitates guiding subsequent agricultural practices and planning. Routine maintenance checks are conducted for post-flight maintenance, which involves inspection of components, cleaning sensors, and ensuring firmware/software updates are applied. Regulatory compliance and safety protocols are also followed to adhere to local aviation regulations governing drone operations. Obtaining necessary permits and licenses for commercial drone usage in agriculture is important. Safety protocols are implemented to mitigate risks, which include maintaining safe distances, avoiding populated areas, and conducting risk assessments. Lastly, the collected data on operational experience is used to refine future missions and improve outcomes. Adaptation to drone settings and protocols based on observed results is mostly carried out. Above all, continuous advancements in drone technology and sensor capabilities further enhance their effectiveness in pinpointing hurdles that have not been anticipated, one such faced by modern agriculture, thus making drones an indispensable tool for sustainable and efficient farming practices.

In contrast to drones used for other applications like surveillance or filmmaking, agricultural drones are specifically tailored for farming needs. They are equipped with specialized sensors that focus on agricultural metrics such as crop health, vegetation indices, and pest infestations. The flight paths and data collection protocols are designed to cover large agricultural fields efficiently and systematically, ensuring comprehensive monitoring and analysis. Data collected by these drones undergoes thorough processing and analysis and anomaly detection to identify issues like water stress or pest outbreaks. This analytical approach enables farmers to make informed decisions regarding irrigation, fertilization, and pest control, thereby optimizing resource use and enhancing crop yields.

2.5 APPLICATIONS OF DRONES IN AGRICULTURE

Drones have revolutionized agricultural practices by offering a wide range of applications that enhance efficiency, productivity, and sustainability across various stages of crop production. Figure 2.4 represents the diverse applications of drones in agriculture.

2.5.1 Crop Monitoring and Assessment

One of the primary applications of drones in agriculture is crop monitoring and assessment. Traditional methods of monitoring crops involve labor-intensive field surveys and manual data collection, which are often time-consuming and limited in scope. Drones, on the other hand, can cover large areas of farmland, allowing farmers to inspect the entire field regularly without the need to physically reach every corner. Drones equipped with RGB cameras, multispectral cameras, and hyperspectral cameras have revolutionized this process by enabling farmers to obtain high-resolution aerial imagery of their fields quickly and cost-effectively. Multispectral cameras capture data across multiple bands, including visible and near-infrared, enabling the calculation of vegetation indices such as Normalized Difference Vegetation Index (NDVI) for precise crop health assessment (Candiago et al., 2015; Khan et al., 2018). By analyzing multispectral imagery captured by drones, farmers can identify areas of their fields experiencing nutrient deficiencies, enabling targeted fertilization interventions to optimize crop yield. Hyperspectral cameras offer an even higher spectral resolution, allowing for the detection of specific biochemical properties in plants and

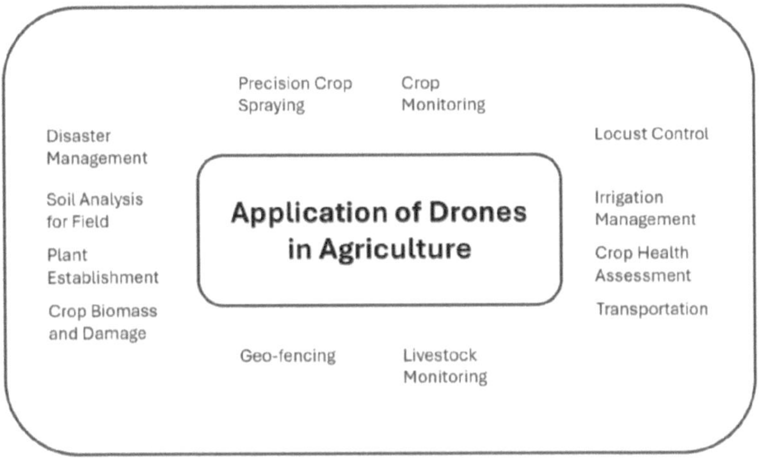

FIGURE 2.4 Applications of drones in agriculture

aiding in early disease detection and nutrient deficiency identification (Ahmad et al., 2021).

Aerial imagery captured by drones provides valuable insights into crop health, growth patterns, and environmental conditions. This data facilitates early detection of stress factors such as nutrient deficiencies, water stress, or pest infestations, allowing farmers to take timely corrective measures and optimize crop yields. Also, by integrating GIS mapping with drones, farmers can draw field borders for accurate flight patterns, manage resources, increase yield, and better manage input costs. Figure 2.5 shows that understanding leaf water content is crucial for determining crop water stress and can be done effectively using drone technology. Several researchers have highlighted the relationship between leaf water content stress and low water potential, indicating an imbalance between evaporated leaf water content and absorbed water level by the root system. Leaf water stress is influenced by plant condition, with transpiration rate and temperature being indirectly related. Higher transpiration rates result in lower crop water stress due to adequate water availability in leaves, whereas low transpiration leads to higher crop water stress.

2.5.2 Precision Spraying and Fertilization

Another significant application of drones in agriculture is precision spraying and fertilization. Traditional methods of applying pesticides and fertilizers often involve blanket spraying of entire fields, resulting in wastage of resources and environmental pollution. Drones equipped with precision spraying systems deliver targeted application of agrochemicals, fertilizers, and pesticides with unparalleled accuracy and efficiency (Mustafi et al., 2021). Sprayers mounted on drones can deliver precise amounts of pesticides and fertilizers to specific areas of the field, reducing chemical drift and minimizing environmental impact. Some drones are also equipped with sensors and imaging technology that enable real-time monitoring of crop health,

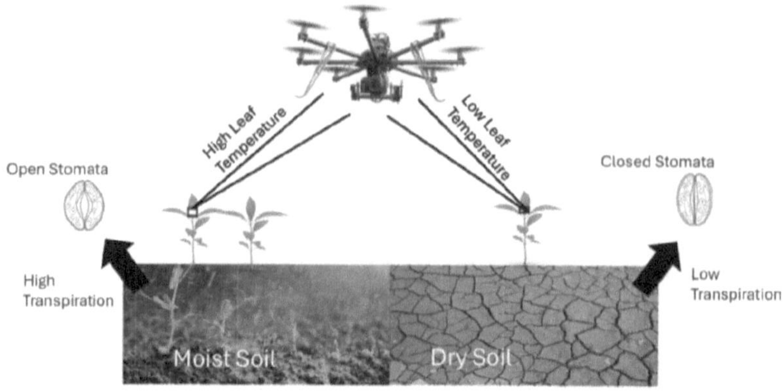

FIGURE 2.5 Assessing crop water stress via drone technology involves analyzing leaf transpiration, temperature, and thermal effects, comparing them with air and soil moisture levels

allowing for on-the-fly adjustments to spraying patterns and chemical concentrations. By precisely targeting specific areas of the field based on real-time data analysis, drones minimize chemical usage, reduce environmental impact, and ensure optimal crop coverage. This approach not only enhances crop protection against pests and diseases but also minimizes input costs for farmers. Figure 2.6 presents how drone-based spraying is preferable over human spraying methods.

Drones equipped with real-time kinematic (RTK) GPS systems enable precise spraying and fertilization, reducing chemical usage and environmental impact (Ahmad et al., 2020; Perez-Ruiz et al., 2021). RTK GPS systems provide centimeter-level accuracy in drone navigation and positioning, ensuring precise application of pesticides and fertilizers to specific locations within the field. Coupled with autonomous flight planning software, RTK GPS technology enables drones to follow predefined spraying patterns with exceptional accuracy, minimizing overlap and reducing chemical usage. For instance, drones equipped with RTK GPS systems can autonomously navigate through fields while adjusting spraying parameters based on real-time data, such as wind speed and direction, ensuring uniform coverage and minimizing environmental impact. The concept of precision agriculture using UAVs is represented in Figure 2.7.

2.5.3 Irrigation Management

Effective irrigation management is critical for optimizing crop yield and conserving water resources. Drones play a crucial role in this aspect of agriculture by providing farmers with accurate data on soil moisture levels, crop water stress, and irrigation efficiency. Thermal cameras mounted on drones can detect variations in surface temperature, allowing farmers to identify areas of the field that are experiencing water stress. This information can be used to adjust irrigation schedules and optimize water usage. LiDAR (Light Detection and Ranging) technology integrated

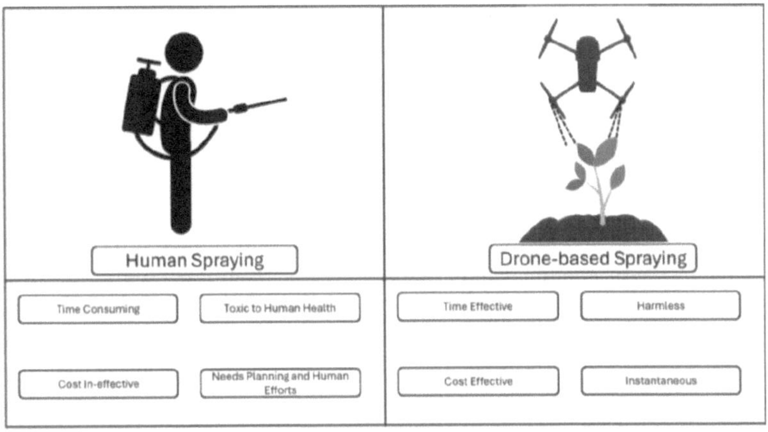

FIGURE 2.6 Human spraying vs. drone-based spraying

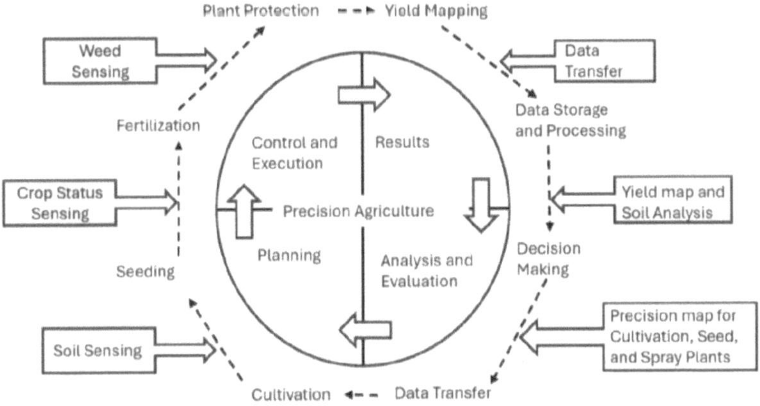

FIGURE 2.7 Concept of precision agriculture using UAVs

into drones facilitates accurate terrain modeling and water flow analysis, optimizing irrigation practices (Kumar et al., 2023; Debnath et al., 2023). Additionally, drones equipped with LiDAR sensors can generate high-resolution topographic maps of fields, enabling farmers to design more efficient irrigation systems based on terrain and soil characteristics. Combined with soil moisture sensors and weather data, LiDAR-derived elevation maps enable precise irrigation scheduling and optimization, reducing water wastage and improving crop water use efficiency. Farmers can utilize LiDAR-generated elevation maps to design variable-rate irrigation systems that adjust water application rates based on slope and soil moisture levels, ensuring optimal hydration across the field.

Drones equipped with thermal imaging and multispectral sensors provide valuable insights into soil moisture levels and crop water stress. By conducting aerial surveys of agricultural fields, drones can identify areas requiring irrigation or drainage adjustments, optimizing water usage and ensuring uniform crop hydration. This proactive approach to irrigation management helps prevent water wastage, mitigate drought stress, and improve overall crop health and yield.

2.6 PEST AND DISEASE DETECTION

Early detection of pests and diseases is essential for preventing crop losses and minimizing the need for chemical interventions. Drones equipped with high-resolution cameras and imaging technology can quickly scan large areas of farmland and identify signs of pest infestation or disease outbreak (Kaivosoja et al., 2021; Hafeez et al., 2022). Multispectral and hyperspectral cameras mounted on drones can detect subtle changes in crop health that may indicate the presence of pests or diseases. Machine learning algorithms can analyze aerial imagery and identify patterns associated with specific pests or diseases, allowing farmers to take proactive measures to mitigate risks.

Machine learning algorithms trained on large datasets of aerial imagery can autonomously detect and classify pests, diseases, and other crop abnormalities with high accuracy. Deep learning models, such as convolutional neural networks (CNNs), can identify subtle visual cues indicative of pest infestations or disease outbreaks, enabling early intervention and targeted management strategies. For example, drones equipped with machine learning-based pest detection systems can patrol farmland autonomously, continuously scanning crops for signs of pest damage or disease symptoms, alerting farmers to potential threats in real time.

2.7 YIELD ESTIMATION AND PREDICTION

Accurate yield estimation is vital for crop planning, resource allocation, and market forecasting. Drones equipped with advanced imaging technology and data analytics software can provide farmers with precise estimates of crop yield well before harvest. By capturing aerial imagery throughout the growing season, drones can monitor crop growth and development over time. Drones, coupled with data analytics and machine learning algorithms, enable accurate estimation and prediction of crop yields. By analyzing aerial imagery, vegetation indices, and other agronomic data, drones can provide insights into crop biomass, canopy coverage, and yield potential. This information assists farmers in making informed decisions related to crop management practices, harvest planning, and resource allocation, ultimately optimizing yield outcomes and maximizing profitability.

Computer vision algorithms analyze aerial imagery to extract crop-related parameters, such as plant count, canopy cover, and biomass density, throughout the growing season (Shahi et al., 2024; Alvarez-Mendoza et al., 2022). Data analytics techniques, including regression analysis and time-series forecasting, are employed to model crop growth dynamics and predict future yield based on historical trends and environmental factors. By aggregating and analyzing drone-collected data over multiple growing seasons, farmers can develop predictive models that forecast crop yield with high accuracy, enabling better decision-making regarding resource allocation, marketing strategies, and financial planning.

Pazhanivelan et al. (2023) used agricultural drones with multispectral and thermal sensors to capture high-resolution imagery over large fields. They focused on vegetation indices like weighted difference vegetation index (density and health of vegetation), visible atmospherically resistant index (chlorophyll concentration), modified chlorophyll absorption ratio index (biomass variations), and normalized green red difference index (early-stage growth patterns). These indices were integrated into regression models predicting crop yield with 76% accuracy ($R^2 = 0.76$). This enabled precise decisions on irrigation, fertilization, pest management, and harvest timing by farmers and practitioners. Another study investigated optimizing drone flight altitudes for estimating *Citrus unshiu* fruit yield. UAV imaging from altitudes of 30 m, 50 m, 70 m, 90 m, and 110 m was analyzed to determine the optimal resolution for accurate fruit size estimation. Histogram equalization of images significantly improved fruit count accuracy compared to untreated images, particularly evident at lower altitudes. For instance, at 30 m altitude, normal images estimated 73, 55, and 88 fruits, while histogram-equalized images estimated 88, 71, and 105, against actual counts of 124, 88, and 141 fruits. Vegetation indices (i.e., integrated principal component analysis) showed comparable accuracy to histogram equalization, but VI I1 showed discrepancies from actual yields. The findings support drones as efficient tools for precise citrus yield estimation, aiding agricultural management with cost-effective, high-resolution data collection methods.

In conclusion, drones have emerged as valuable tools for modern agriculture, offering a wide range of applications that contribute to increased productivity, efficiency, and sustainability. From crop monitoring and assessment to precision spraying and fertilization, irrigation management, pest and disease detection, and yield estimation and prediction, drones are revolutionizing farming practices around the world. As technology continues to advance and drones become more accessible to farmers of all scales, the agricultural industry stands to benefit significantly from their continued integration into everyday operations.

2.8 ADVANTAGES OF DRONES IN AGRICULTURE

In recent years, the integration of drones, or Unmanned Aerial Systems (UAS), into agricultural practices has ushered in a new era of precision farming. With step-up of advanced technologies such as sensors, cameras, and data analytics, drones offer a range of scientific and accurate benefits that revolutionize traditional agricultural methods. From crop monitoring to pest management and yield prediction, drones are proving to be invaluable tools for enhancing efficiency, sustainability, and productivity in agriculture (Figure 2.8).

2.8.1 Precision Agriculture

At the forefront of drone applications in agriculture is precision farming. Drones equipped with high-resolution cameras, multispectral sensors, and LiDAR

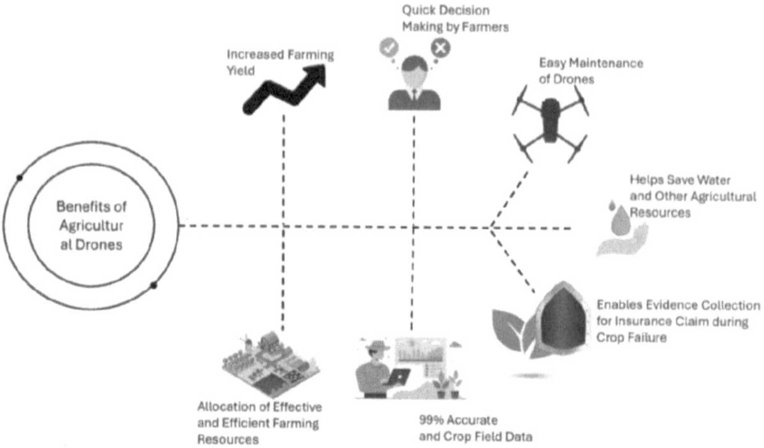

FIGURE 2.8 Advantages of drones in agriculture

technology enable farmers to gather detailed information about their fields with unparalleled precision (Kumar et al., 2023). By capturing aerial imagery and data, drones provide valuable insights into crop health, nutrient levels, moisture content, and pest infestations. This data allows farmers to make informed decisions regarding irrigation, fertilization, and pest control, optimizing resource allocation and minimizing environmental impact.

2.8.2 Enhanced Crop Monitoring

One of the primary benefits of drones in agriculture is their ability to monitor crops more efficiently than traditional methods. With drones, farmers can survey large areas of farmland quickly and regularly, detecting early signs of stress, disease, or pest infestation. Multispectral and hyperspectral imaging capabilities allow drones to identify subtle variations in plant health and growth, enabling targeted interventions to address issues before they escalate (Sarvakar et al., 2024). By monitoring crops with precision and accuracy, drones help farmers maximize yield potential while minimizing input costs.

2.8.3 Real-Time Data Collection

Drones provide real-time access to critical agricultural data, enabling farmers to respond promptly to changing conditions in the field. Equipped with onboard sensors and GPS technology, drones can collect and transmit data instantaneously, allowing farmers to make time-sensitive decisions regarding crop management (Pathak et al., 2018). Whether adjusting irrigation schedules, deploying pest control measures, or assessing crop damage after extreme weather events, the ability to access real-time data enhances operational efficiency and reduces reliance on subjective observations.

2.8.4 Sustainable Resource Management

By enabling precise targeting of inputs such as water, fertilizers, and pesticides, drones promote sustainable resource management in agriculture. Rather than applying these inputs uniformly across entire fields, drones allow farmers to apply them only where and when they are needed most. This targeted approach minimizes waste, reduces environmental pollution, and conserves resources, contributing to a more sustainable agricultural system (Yadav et al., 2023). In addition to that, by optimizing resource use and minimizing input costs, drones help improve the economic viability of farming operations.

2.8.5 Improved Crop Health Monitoring

Drones equipped with advanced sensors and imaging technology offer unparalleled capabilities for monitoring crop health and identifying potential issues early on. Multispectral and thermal imaging allow drones to detect subtle changes in plant physiology, such as water stress, nutrient deficiencies, or disease symptoms, that may not be visible to the naked eye (Shukla et al., 2021). By identifying these issues in their early stages, farmers can take proactive measures to mitigate risks and prevent yield losses, ultimately improving crop health and resilience.

2.8.6 Accurate Pest Management

Pest management is a critical aspect of agricultural production, and drones are proving to be invaluable tools in this regard. Equipped with high-resolution cameras and machine learning algorithms, drones can autonomously detect and identify pest infestations in crops with remarkable accuracy lost (Filho et al., 2020). By precisely mapping the extent of infestations and identifying affected areas, drones enable targeted application of pest control measures, reducing the need for broad-spectrum pesticides and minimizing environmental harm. This targeted approach not only improves pest control efficacy but also helps preserve beneficial insect populations and biodiversity.

2.8.7 Data-Driven Decision-Making

Perhaps the most significant benefit of drones in agriculture is their ability to facilitate data-driven decision-making. By collecting, analyzing, and visualizing vast amounts of data from the field, drones provide farmers with actionable insights that inform strategic decisions. Whether optimizing planting schedules, adjusting irrigation regimes, or selecting optimal harvesting times, data-driven decision-making maximizes productivity, minimizes risks, and enhances overall farm management (Paul et al., 2022). Furthermore, by leveraging historical data and predictive analytics, drones empower farmers to anticipate future challenges and opportunities, enabling proactive planning and adaptation to changing environmental conditions.

In summary, the integration of drones into agricultural practices offers a myriad of scientific and accurate benefits that enhance efficiency, sustainability, and productivity. From precision agriculture and enhanced crop monitoring to sustainable resource management and data-driven decision-making, drones are transforming traditional farming methods and revolutionizing the way food is produced. As drone technology continues to advance and become more accessible, its role in agriculture is expected to expand further, driving innovation and positive change across the agricultural sector.

2.9 CASE STUDY

2.9.1 Case Study 1: Detection of *Rhynchophorus ferrugineus* Infestation in Palm Trees Using Drones

Psirofonia et al. (2017) conducted a case study on the devastating impact of the red palm weevil *R. ferrugineus* on palm tree plantations, particularly in regions across Asia and the Mediterranean basin, where it has been confirmed in over 30 countries. This destructive pest caused severe damage to palm trees by infesting them, leading to complete leaf loss, trunk rot, and eventual death. The life cycle of the weevil involved adult females laying eggs in the crown of palm trees, with larvae subsequently tunneling through the upper trunk, hollowing it out, and causing irreparable harm. Human activities, such as the transportation of infested palm trees and offshoots, contributed significantly to the rapid spread of this pest.

In the specific context of Crete, the presence of *R. ferrugineus* posed a significant threat to the Cretan date palm *Phoenix theophrasti*, which was found in the palm tree forest of Vai, the last palm tree forest on Earth containing this endangered plant species. Given the importance of early detection in mitigating the spread of the pest and minimizing damage, efforts were directed toward using Unmanned Aerial Vehicles (UAVs) for inspection purposes.

By employing drones equipped with specialized sensors and imaging technology, authorities were able to conduct aerial surveys of palm tree plantations to detect early signs of infestation. These UAVs could cover large areas quickly and efficiently, allowing for timely identification of infested trees. Once infested trees were located, they were targeted for immediate destruction to prevent further spread of the pest. Additionally, endangered trees were treated with insecticides to protect them from infestation, thus safeguarding the overall health of the palm tree population.

In summary, the case study highlighted the critical role of UAV technology in early detection and management of *R. ferrugineus* infestations in palm tree plantations. By leveraging drones for aerial surveillance, authorities were able to implement proactive measures to control the spread of the pest and preserve valuable palm tree populations, such as the endangered Cretan date palm in the Vai Forest of Crete.

2.9.2 Case Study 2: Enhancing Sugar Beet Disease Management Through Drone-Based Image Processing

Altas et al. (2018) conducted a case study using image processing techniques with drones to determine the severity of sugar beet leaf spot disease (*Cercospora beticola* Sacc.) in the vicinity of Tokat province, Turkey, in a local sugar beet field. They utilized a DJI Phantom 3 Advanced drone equipped with a 12-megapixel camera capable of capturing images with up to 1080P resolution. The drone featured a GPS positioning system, allowing precise image capture and location tracking. The study compared disease severity assessments visually by experts with those obtained through image processing algorithms. Image processing techniques were employed using MATLAB version R2014a, leveraging the Image Processing Toolbox module for analysis.

The study demonstrated that image processing techniques, when coupled with drone technology, offer a precise and efficient method for assessing disease severity in sugar beet plants. This method helped farmers by providing a faster and more accurate means of disease detection, enabling timely intervention and prevention of yield losses. It reduced labor and time costs associated with traditional visual assessments, thereby enhancing agricultural productivity and sustainability.

2.10 CURRENT STATUS OF DRONE USAGE IN INDIAN AGRICULTURE

Several drone-based agricultural projects are currently operational in India. Initially, the Indian government sanctioned the use of drones for agricultural research by institutions like the International Crops Research Institute for the Semi-Arid Tropics (ICRISAT) in Hyderabad. This move aimed to encourage emerging researchers and entrepreneurs to develop affordable drone solutions for India's vast farming community.

Collaborations between drone startups and agricultural corporations have been instrumental in leveraging drone technology for precision agriculture in India. For instance, a Bengaluru-based startup partnered with Syngenta, a Swiss agri-business company, to analyze suitable corn cultivation regions in India using drones. Furthermore, drones have been employed in various agricultural activities across the country. In Kerala, drones developed by startups were utilized to spray micronutrients over paddy fields in Alathur, Kochi, demonstrating their potential for crop management. In regions like Rajasthan, drones were deployed to combat locust infestations by efficiently spraying insecticides over affected areas.

Several state agricultural universities, such as Acharya NG Ranga Agriculture University in Andhra Pradesh, have initiated plans to incorporate drones for pesticide spraying and other agricultural tasks. These universities have acquired drones for testing and plan to expand their usage pending successful trials. Government bodies like the Ministry of Civil Aviation (MoCA) and the Directorate General of Civil Aviation (DGCA) have granted conditional exemptions to organizations for drone operations under the Unmanned Aircraft System (UAS) Rules, 2021. These exemptions aim to facilitate the integration of drone technology into agricultural practices. Moreover, private companies like Tractors and Farm Equipment Limited (Chennai), Mahindra & Mahindra (Mumbai), and Bayer Crop Science (Mumbai) are conducting agricultural trials using drones for tasks such as aerial spraying, crop health monitoring, and disease prevention.

In conclusion, the use of agricultural drones holds immense potential to revolutionize traditional farming methods in India. While mastering drone technology may pose initial challenges for farmers, the benefits in terms of enhanced productivity and data-driven decision-making are substantial. As drones continue to evolve and become more accessible, their significance in Indian agriculture is expected to grow significantly in the coming years.

2.11 CHALLENGES IN THE IMPLEMENTATION OF DRONES FOR AGRICULTURAL USE

Implementing drone technology in agriculture holds tremendous potential to revolutionize farming practices, offering benefits like precision monitoring, efficient resource management, and enhanced productivity. However, this promising technology also confronts significant challenges that hinder its widespread adoption and optimal utilization across various aspects of agriculture given as follows:

1. Flight time and range limitations

One of the significant challenges hindering the efficiency of agricultural drones is their limited flight time and range, typically operating between 20 and 60 minutes per charge (Zhang & Kovacs, 2012). This constraint, a result of payload and battery capacity, forces farmers to plan meticulously to maximize operational efficiency. Given that crop monitoring often necessitates covering extensive areas, these flight constraints impose practical difficulties, especially for those managing diverse crop rotations (Xiang & Tian, 2011).

2. Initial cost and return on investment

The high initial cost is another substantial barrier to drone adoption in agriculture. Advanced agricultural drones can cost upwards of $25,000, encompassing expenses for the drone, software, hardware, and necessary training. For small and medium-sized farms, this investment represents a significant financial challenge, with long-term return on investment often uncertain. Studies indicate that while drones can improve productivity and reduce costs in the long run, the upfront expense remains a deterrent.

3. Regulatory complexities

Farmers often face regulatory challenges when integrating drones into agricultural practices. National aviation authorities, such as India's Directorate General of Civil Aviation, impose strict regulations requiring permits, adherence to flight restrictions, and compliance with safety protocols (Pathak et al., 2018). Navigating these regulations can prove both time-consuming and complex, discouraging farmers from adopting drone technology.

4. Connectivity and data handling

In remote agricultural areas, inadequate connectivity presents significant challenges for effective drone operations. Real-time data transmission necessary for live monitoring and GPS navigation often relies on internet coverage, which may be poor in these regions (Bendig et al., 2014). Alternative solutions, such as drones with onboard data storage or satellite-based communication systems, are required but can further increase costs and complexity, limiting accessibility for smaller farms (Colomina & Molina, 2014).

5. Weather dependency and operational reliability

Drones are highly sensitive to adverse weather conditions, such as wind and precipitation, which can disrupt operations and compromise data integrity. Unlike traditional aircraft, drones cannot operate in poor weather, limiting their reliability for consistent agricultural monitoring and surveillance. This weather dependency underscores the need for robust, all-weather-capable drones (Hunt et al., 2010).

6. Skill and knowledge requirements

Effective use of drone technology in agriculture requires specialized skills in drone piloting, data analysis, and software interpretation. Farmers and agronomists must be adept in these areas to use drones safely and translate aerial imagery into actionable farm management insights (Puri et al., 2017). The need for specialized training and expertise presents an additional barrier to widespread adoption.

7. Privacy and ethical considerations

The potential misuse of drones concerning privacy infringements and unauthorized data collection is a growing concern among farmers and communities. Strict adherence to privacy regulations and ethical guidelines is crucial to ensure responsible drone use and maintain public trust (Clarke, 2014). Addressing these concerns is imperative to foster a positive perception and acceptance of agricultural drones.

The challenges are not limited to this but also extend to atmospheric variability, radiometric calibration, and geometric distortions that might affect the accuracy, reliability, and interpretability of data collected by drones, influencing decisions related to crop management, yield prediction, and resource allocation (Delavarpour et al., 2021).

Atmospheric conditions like humidity and aerosols affect how sensors capture light, impacting image clarity and spectral accuracy. Turbulence caused by wind can lead to image distortions and compromise spatial resolution, complicating precise mapping efforts (Delavarpour et al., 2021). Radiometric challenges arise from variations in light intensity due to solar elevation changes, atmospheric conditions, and cloud cover, which introduce errors in data comparison. Geometric challenges related to sensor positioning, platform motion, and terrain effects lead to spatial inaccuracies in captured images. Mitigating these issues involves integrating meteorological insights into flight planning and employing advanced correction methods such as radiometric and geometric calibration (Colomina & Molina, 2014).

Despite the formidable challenges impeding the widespread adoption of drones in agriculture, their potential to revolutionize farming practices is undeniable. Addressing constraints such as limited flight time, high initial costs, regulatory complexities, connectivity issues, weather dependency, skill requirements, and privacy concerns can pave the way for a more efficient and sustainable agricultural future. Continuous technological advancements and adaptive regulatory frameworks promise an integral role for drones in modern farming operations.

2.12 FUTURE SCOPE

The agricultural sector has long grappled with significant challenges such as lack of irrigation infrastructure, temperature fluctuations, groundwater depletion, food scarcity, and wastage, among others. The future of farming largely hinges on the adoption of various cognitive solutions. While extensive research is ongoing and some applications are already available, the industry remains largely underserved. Finding out real-world challenges faced by farmers and utilizing autonomous decision-making and predictive solutions to resolve them is still in its infancy. To fully tap into the vast potential of AI in agriculture, applications must become more robust. This requires the ability to adapt to frequent changes in external conditions, enabling real-time decision-making and employing suitable frameworks/platforms for efficient collection of contextual data.

The high cost of existing cognitive solutions poses a barrier to widespread adoption in farming. Making these solutions more affordable is essential to ensure widespread access to the technology. An open-source platform could significantly lower costs, leading to rapid adoption and greater penetration among farmers. This technology holds promise in enhancing crop yields and ensuring consistent seasonal harvests, particularly in regions like India where agriculture heavily relies on monsoon patterns. AI can aid in predicting weather conditions and other agricultural factors such as soil quality, groundwater levels, crop cycles, and pest outbreaks. Accurate projections facilitated by AI can alleviate many concerns faced by farmers.

AI-driven sensors play a crucial role in collecting vital agricultural data, which can be leveraged to improve production efficiency. These sensors have immense potential in agriculture, enabling scientists to gather data on soil quality, weather patterns, groundwater levels, and more, ultimately enhancing the cultivation process. AI-powered sensors can be integrated into robotic harvesting equipment to optimize data collection. It is estimated that AI-based advisories could potentially increase production by up to 30%.

One of the most significant challenges in farming is crop damage caused by various disasters, including pest infestations. Often, farmers lack timely information to protect their crops effectively. In the digital age, technology can play a vital role in safeguarding crops. AI-enabled image recognition, for instance, can aid in identifying pest attacks. Many companies have successfully implemented drones for monitoring crop production and detecting pests, showcasing the potential of such technology in crop protection.

An illustrative example of AI's transformative potential in agriculture is its application at NatureFresh Farms, where AI algorithms predict the ripening time of tomatoes based on images of their blossoms. This technology streamlines operations and enhances profitability, highlighting AI's role in revolutionizing agriculture. With the global population projected to reach nine billion by 2050, coupled with the challenges posed by climate change, there is an urgent need to increase food production. AI presents a promising solution to this challenge, offering insights into hybrid cultivations and optimizing resource utilization.

A recent study investigated IoT-enabled drones for precision farming to spray insecticides accurately on crop green zones, demonstrated in a grape field near Nashik, Maharashtra, India (Borah et al., 2023). They designed a 16-liter drone that would cover 1 acre in 7–10 minutes, reducing labor costs by 65%, spraying time by 85%, and insecticide use by 50%. Synchronizing the drone's camera and spraying mechanism through IoT platforms would further optimize efficiency. Time-series analysis of drone-captured images and videos can predict crop yields and support machine learning algorithms for disease forecasting and stress management. Shah et al. (2023) demonstrated the effectiveness of convolutional neural network-based deep learning techniques for detecting plant diseases from images captured by drones. They analyzed high-resolution drone-captured images of crop leaves using a trained EfficientNet-B3 model, achieving an impressive accuracy rate of 98.80% in identifying various plant diseases and their corresponding conditions.

In conclusion, AI is poised to revolutionize agriculture in the coming years, offering solutions to enhance productivity, optimize resource management, and address food security concerns. However, realizing this potential requires overcoming challenges such as cost barriers, data collection limitations, and technological integration. Nonetheless, with concerted efforts and advancements in AI technology, agriculture stands to benefit significantly from digital transformation, paving the way for a more sustainable and efficient food production system.

2.13 CONCLUSIONS

The adoption of drones in agriculture represents a significant advancement in modern farming practices, offering a myriad of benefits that enhance efficiency, sustainability, and productivity. From precision agriculture and enhanced crop monitoring to sustainable resource management and data-driven decision-making, drones have the potential to revolutionize traditional farming methods and address key challenges facing the agricultural sector. Case studies showcasing the practical applications of drones, such as the detection of pest infestations in palm tree plantations and the management of sugar beet diseases, highlight the tangible benefits of drone technology in improving crop health and yield outcomes. A point to be noted is that the current status of drone usage in Indian agriculture demonstrates the growing momentum toward integrating drones into farming operations to enhance productivity and address agricultural challenges. Though several challenges hinder the widespread adoption of drones in agriculture, including limited battery life, high costs, regulatory requirements, and safety concerns, there is a bright side to look at with the potential technology and research. A meticulous understanding of these challenges is crucial to unlocking the full potential of drone technology in agriculture and ensuring its accessibility to farmers of all scales. Looking ahead, the future scope of drone technology in agriculture is promising, with opportunities for further innovation and advancement. AI-driven solutions, open-source platforms, and sensor technology hold the key to overcoming existing challenges and maximizing the benefits of drone technology in farming. To conclude, drones have emerged as invaluable tools for modern agriculture, offering solutions to enhance productivity, optimize resource management, and address food security concerns. With continued advancements and strategic interventions, drone technology has the potential to transform the agricultural landscape, paving the way for a more sustainable and efficient food production system.

REFERENCES

Ahmad, A., Ordoñez, J., Cartujo, P., & Martos, V. (2020). Remotely piloted aircraft (RPA) in agriculture: A pursuit of sustainability. *Agronomy*, 11(1), 7.

Ahmad, U., Alvino, A., & Marino, S. (2021). A review of crop water stress assessment using remote sensing. *Remote Sensing*, 13(20), 4155.

Alvarez-Mendoza, C. I., Guzman, D., Casas, J., Bastidas, M., Polanco, J., Valencia-Ortiz, M., Montenegro, F., Arango, J., Ishitani, M., & Selvaraj, M. G. (2022). Predictive modeling of above-ground biomass in brachiaria pastures from satellite and UAV imagery using machine learning approaches. *Remote Sensing*, 14(22), p.5870.

Altaş, Z., Özgüven, M. M., & Yanar, Y. (2018). "Determination of Sugar Beet Leaf Spot Disease Level (Cercospora beticola Sacc.) with Image Processing Technique by Using Drone." *Current Investigations in Agriculture and Current Research*, 5(3), pp. 621–631. DOI: 10.32474/CIACR.2018.05.000214

Bendig, J., Bolten, A., Bennertz, S., Broscheit, J., Eichfuss, S., & Bareth, G. (2014). Estimating biomass of barley using crop surface models (CSMs) derived from UAV-based RGB imaging. *Remote Sensing*, 6(11), 10395–10412.

Borah Crabb, M., Wright, P., Humphrey, O., Johnson, G., Rush, S., van Rein, H., & Hinchen, H. (2019). Unmanned Aerial Vehicles for use in marine monitoring – Marine Monitoring Platform Guidelines No. 3. Joint Nature Conservation Committee (JNCC), Peterborough. ISSN 2517-7605. Available at: https://hub.jncc.gov.uk/assets/3b1a059f-5c48-493b-8002-ff2f68276b15

Brown, A. X., & Johnson, E. N. (2011). Modeling and Flight Testing of the Longitudinal Dynamics of an Inflatable Wing UAV. AIAA Atmospheric Flight Mechanics Conference, Portland, Oregon, August 8–11, 2011. https://doi.org/10.2514/6.2011-6327

Candiago, S., Remondino, F., De Giglio, M., Dubbini, M., & Gattelli, M. (2015). Evaluating multispectral images and vegetation indices for precision farming applications from UAV images. *Remote Sensing*, 7(4), 4026–4047.

Clarke, R. (2014). Understanding the drone epidemic. *Computer Law and Security Review*, 30(3), 230–246.

Colomina, I., & Molina, P. (2014). Unmanned aerial systems for photogrammetry and remote sensing: A review. *ISPRS Journal of Photogrammetry and Remote Sensing*, 92, 79–97.

Debnath, S., Paul, M., & Debnath, T. (2023). Applications of LiDAR in agriculture and future research directions. *Journal of Imaging*, 9(3), 57.

Delavarpour, N., Koparan, C., Nowatzki, J., Bajwa, S., & Sun, X. (2021). A technical study on UAV characteristics for precision agriculture applications and associated practical challenges. *Remote Sensing*, 13(6), 1204.

García, Fernández, María, Yuri Álvarez López, & Fernando Luis Las Heras Andrés. (2020). "3D-SAR Processing of UAV-Mounted GPR Measurements: Dealing with Non-Uniform Sampling." European Conference on Antennas and Propagation (EuCAP), pp. 9135650. https://doi.org/10.23919/EuCAP48036.2020.9135650.

Greenfield, J., & Thaxton, C. (2022). "Data Acquisition and Logging System for a Drone-mounted Atmospheric Sensor Suite," SoutheastCon 2022, Mobile, AL, USA, 2022, pp. 178–183, doi: 10.1109/SoutheastCon48659.2022.9764113.

Hafeez, A., Husain, M. A., Singh, S. P., Chauhan, A., Khan, M. T., Kumar, N., Chauhan, A., & Soni, S. K. (2022). Implementation of drone technology for farm monitoring & pesticide spraying: A review. *Information processing in Agriculture*, 10(2), 192–203.

Hunt Jr., E. R., Hively, W. D., Fujikawa, S. J., Linden, D. S., Daughtry, C. S., & McCarty, G. W. (2010). Acquisition of NIR-green-blue digital photographs from unmanned aircraft for crop monitoring. *Remote Sensing*, 2(1), 290–305.

Filho, F. H., Heldens, W. B., Kong, Z., & de Lange, E. S. (2020). Drones: innovative technology for use in precision pest management. *Journal of Economic Entomology*, 113(1), 1–25.

Javaid, M., Haleem, A., Singh, R. P., & Suman, R. (2022). Enhancing smart farming through the applications of Agriculture 4.0 technologies. *International Journal of Intelligent Networks*, 3, 150–164.

Kaivosoja, J., Hautsalo, J., Heikkinen, J., Hiltunen, L., Ruuttunen, P., Näsi, R., Niemeläinen, O., Lemsalu, M., Honkavaara, E., & Salonen, J. (2021). Reference measurements in developing UAV systems for detecting pests, weeds, and diseases. *Remote Sensing*, 13(7), 1238.

Kangunde, V., Jamisola Jr., R. S., & Theophilus, E. K. (2021). A review on drones controlled in real-time. *International Journal of Dynamics and Control*, 9(4), 1832–1846.

Khan, Z., Rahimi-Eichi, V., Haefele, S., Garnett, T., & Miklavcic, S. J. (2018). Estimation of vegetation indices for high-throughput phenotyping of wheat using aerial imaging. *Plant Methods*, 14, 1–11.

Kumar, S. P., Subeesh, A., Jyoti, B., & Mehta, C. R. (2023). Applications of Drones in Smart Agriculture. In *Smart Agriculture for Developing Nations: Status, Perspectives and Challenges* (pp. 33–48). Singapore: Springer Nature Singapore.

Mustafi, S., Ghosh, P., Roy, K., Dan, S., Mukherjee, K., & Nath Mandal, S. (2021). Drones for intelligent agricultural management. In *IoT-based Intelligent Modelling for Environmental and Ecological Engineering: IoT Next Generation EcoAgro Systems* (pp. 81–100). Cham: Springer International Publishing.

Otto, A., Agatz, N. A. H., Campbell, J. F., Golden, B., & Pesch, E. (2018). 'Optimization approaches for civil applications of unmanned aerial vehicles (UAVs) or aerial drones: A survey.' *Networks*, 72, 411–58.

Pathak, R., Barzin, R., & Bora, G. C. (2018). Data-driven precision agricultural applications using field sensors and Unmanned Aerial Vehicle. International Journal of Precision Agricultural Aviation, 1(1), 19–23.

Paul, K., Chatterjee, S. S., Pai, P., Varshney, A., Juikar, S., Prasad, V., Bhadra, B., & Dasgupta, S. (2022). Viable smart sensors and their application in data driven agriculture. *Computers and Electronics in Agriculture*, 198, 107096.

Pazhanivelan, S., Kumaraperumal, R., Shanmugapriya, P., Sudarmanian, N. S., Sivamurugan, A. P., & Satheesh, S. (2023). Quantification of biophysical parameters and economic yield in cotton and rice using drone technology. *Agriculture*, 13(9), 1668.

Perez-Ruiz, M., Martínez-Guanter, J., & Upadhyaya, S. K. (2021). High-precision GNSS for agricultural operations. In *GPS and GNSS Technology in Geosciences* (pp. 299–335). Amsterdam: Elsevier.

Psirofonia, P., Samaritakis, V., Eliopoulos, P., & Potamitis, I. (2017). Use of unmanned aerial vehicles for agricultural applications with emphasis on crop protection: Three novel case-studies. *International Journal of Agricultural Science and Technology*, 5(1), 30–39.

Puri, V., Nayyar, A., & Raja, L. (2017). Agriculture drones: A modern breakthrough in precision agriculture. *Journal of Statistics and Management Systems*, 20(4), 507–518.

Radoglou-Grammatikis, P., Sarigiannidis, P., Lagkas, T., & Moscholios, I. (2020). A compilation of UAV applications for precision agriculture. *Computer Networks*, 172, 107148.

Rejeb, A., Abdollahi, A., Rejeb, K., & Treiblmaier, H. (2022). Drones in agriculture: A review and bibliometric analysis. *Computers and Electronics in Agriculture*, 198, 107017.

Sankhua, R. N. (2022). Rejuvenating Water Wisdom: A Route to Resilience. In *Sustainability of Water Resources: Impacts and Management* (pp. 79–94). Cham: Springer International Publishing.

Sarvakar, K., & Thakkar, M. (2024). Different Vegetation Indices Measurement Using Computer Vision. In *Applications of Computer Vision and Drone Technology in Agriculture 4.0* (pp. 133–163). Singapore: Springer Nature Singapore.

Shahi, T. B., Khadka, R. B., & Neupane, A. (2024). Applicability of UAV in Crop Health Monitoring Using Machine Learning Techniques. In Recent Advances in Machine Learning for Crop Monitoring (IGI Global).

Shahi, Tej & Xu, Cheng-Yuan & Neupane, Arjun & Guo, William. (2023). Recent Advances in Crop Disease Detection Using UAV and Deep Learning Techniques. *Remote Sensing*, 15(9), 2450.

Shakhatreh, H., Sawalmeh, A. H., Al-Fuqaha, A., Dou, Z., Almaita, E., Khalil, I., Othman, N. S., Khreishah, A., & Guizani, M. (2019). Unmanned aerial vehicles (UAVs): A survey on civil applications and key research challenges. *IEEE Access*, 7, 48572–48634.

Shukla, R., Dubey, G., Malik, P., Sindhwani, N., Anand, R., Dahiya, A., & Yadav, V. (2021). Detecting crop health using machine learning techniques in smart agriculture system. *Journal of Scientific & Industrial Research*, 80(08), 699–706.

Talaviya, T., Shah, D., Patel, N., Yagnik, H., & Shah, M. (2020). Implementation of artificial intelligence in agriculture for optimisation of irrigation and application of pesticides and herbicides. *Artificial Intelligence in Agriculture*, 4, 58–73.

Villa, T. F., Gonzalez, F., Miljievic, B., Ristovski, Z. D., & Morawska, L. (2016). An overview of small unmanned aerial vehicles for air quality measurements: Present applications and future prospectives. *Sensors*, 16(7), 1072.

Xiang, H., & Tian, L. (2011). Development of a low-cost agricultural remote sensing system based on an autonomous unmanned aerial vehicle (UAV). *Biosystems Engineering*, 108(2), 174–190.

Yadav, N., & Sidana, N. (2023). Precision agriculture technologies: Analysing the use of advanced technologies, such as drones, sensors, and gps, in precision agriculture for optimizing resource management, crop monitoring, and yield prediction. *Journal of Advanced Zoology*, 44.

Zhang, C., & Kovacs, J. M. (2012). The application of small unmanned aerial systems for precision agriculture: A review. *Precision Agriculture*, 13, 693–712.

3 Drone Swarms and Autonomous Systems

Velmurugan S, Sudharsan K K, Sri Sathya Priya R, Ruthresh T, Sivakumar P, and Ramesh A

3.1 INTRODUCTION

The field of aerial operations is changing due to the convergence of artificial intelligence (AI), data science, and cybersecurity, especially with regard to autonomous and swarm drones. This evolution necessitates a comprehensive strategy that addresses cybersecurity issues and makes use of AI and data science to improve capabilities. While this connection promises unmatched efficiency, it also emphasizes how crucial it is to keep these systems safe from hostile attacks. In this chapter, we will discuss the critical role cybersecurity plays in drone operations as well as the revolutionary potential of AI and data science to spur innovation in this field.

3.2 CYBERSECURITY FOR DRONE AND UNMANNED AERIAL VEHICLES

3.2.1 Introduction

Cybersecurity, a critical field in the current digital era, is the process of preventing unwanted access to computer networks, systems, and data, cyberattacks, and breaches. As technology develops and permeates more areas of our lives, cybersecurity becomes increasingly important for protecting private data and guaranteeing the availability, integrity, and confidentiality of digital assets.

3.2.2 Cybersecurity in UAV

When we consider the application of cybersecurity in drones, also known as unmanned aerial vehicles (UAVs), it becomes particularly crucial due to the unique vulnerabilities and risks associated with these unmanned aircraft. UAVs rely heavily on digital systems, including communication networks, onboard computers, and navigation systems, making them susceptible to cyber threats such as hacking, data interception, and remote-control hijacking.

DOI: 10.1201/9781003481584-3

3.2.3 Terminologies

In the fields of cybersecurity and UAVs, various terminologies are utilized, which include:

CIA Triad: Representing confidentiality, integrity, and availability, the CIA Triad serves as the foundation for creating security policies and procedures. Integrity and confidentiality guard against unwanted access to private data, and availability ensures timely access to information for authorized users.

STRIDE Model: One well-known approach for recognizing and classifying computer security threats is the STRIDE model. Originally introduced by Microsoft, it offers a structured approach to understanding various types of security vulnerabilities and their potential impact on system integrity. The model identifies six categories of threats, each of which represents a different way in which a system's security can be compromised:

Spoofing: Spoofing occurs when a malicious actor impersonates a legitimate entity or system component to gain unauthorized access or privileges. This can include spoofing IP addresses, email addresses, or user identities to get around security measures and obtain illicit access to confidential data or resources. The integrity of systems is seriously threatened by spoofing assaults, which can lead to various security breaches if left unchecked.

Tampering: Tampering is referred to as unauthorized data modification or alteration of software or system components with malicious intent. This can involve altering the content of files or messages, modifying configuration settings, or injecting malicious code into software or firmware. Tampering attacks can undermine the integrity of a system by introducing errors, vulnerabilities, or backdoors that can be exploited by attackers to achieve their objectives.

Repudiation: Repudiation refers to the ability of an attacker to deny or refute their actions or transactions within a system. This can include denying responsibility for unauthorized access or activities, falsifying audit logs or records, or disputing the authenticity of communications or transactions.

Information Disclosure: Information disclosure involves the unauthorized exposure or leakage of sensitive or confidential information to unauthorized parties. This can happen in a number of ways, such as illegal access to databases or file systems, eavesdropping, or data breaches. Information disclosure attacks can result in the exposure of sensitive personal or corporate data, leading to privacy violations, identity theft, or financial fraud.

Denial of Service (DoS): This attack comprises the deliberate disruption or degradation of service availability by overwhelming a system or network with excessive traffic or requests. This can include flooding a server with network packets, exhausting system resources, or exploiting vulnerabilities to crash or disable critical services.

Elevation of Privilege: Elevation of privilege attacks involve exploiting vulnerabilities in a system or application to escalate the privileges of an unauthorized user or process. This can include bypassing access controls, exploiting privilege escalation vulnerabilities, or abusing misconfigurations to gain elevated access rights. Elevation of privilege attacks can enable attackers to bypass security controls, access sensitive information, or run malicious programs with high privileges, which puts system security at serious risk (Shafik, Matinkhah, & Shokoor, 2023).

3.2.4 Existing Threats and Vulnerabilities

UAVs and drones face numerous threats and vulnerabilities, including design flaws, lack of wireless security, and absence of video encryption. Malicious actors exploit these vulnerabilities using various hacking strategies, such as password theft, man-in-the-middle (MITM) attacks, and distributed denial-of-service (DDoS) attacks. Password theft involves stealing the credentials of the drone operator or the Ground Station (GS), man-in-the-middle attacks involve listening in on and changing the conversation between the GS and the drone, and DDoS attacks involve flooding the drone or the GS with excessive requests or traffic, overwhelming their capacity and disrupting their functionality.

Attacks Done by Password Cracking: Password-related cyberattacks have revealed a rise in linked devices. As a basic illustration of UAV control and revealed positive security weaknesses, password cracking in commercial UAVs is an authentication incursion. Violations in the intercommunication method allow the attackers to get access to the device and manipulate the robot's operating scheme's control mechanisms.

Parrot Bebop 2 and DJI Phantom 4: Drones, including the Parrot Bebop 2 and DJI Phantom 4 Pro, have been targets of several cyberattacks. Two particular dangers to the Phantom 4 Pro include GPS spoofing and the DJI software development kit (SDK).

GPS Spoofing: From Figure 3.1, the GPS is not encrypted when it comes to phantom models. Furthermore, setting a private GPS is simple. Through the use of a fake GPS code exchanged with the UAV flight controller, the intruder deceives the opponent's GPS parody and breaks through the UAV defenses. If the attacker is successful, the UAV is still entirely within their hands. One might use the LabSat3 GPS simulator to conduct this assault on a UAV without a spoof detection device.

DoS Attacks: Through this method, two network devices can be disconnected while utilizing all of their available bandwidth. Ruining a specific service and blocking access is the aim. In order to stop authorized users from doing so, it usually consumes up all of the device's bandwidth or overloads its processing capacity.

Parrot AR Drone 2.0 and Cheerson CX-10W: A survey was conducted using the Cheerson CX-10W and the Parrot AR Drone 2.0; other hacking methods were tested separately. There were six recorded successful attacks.

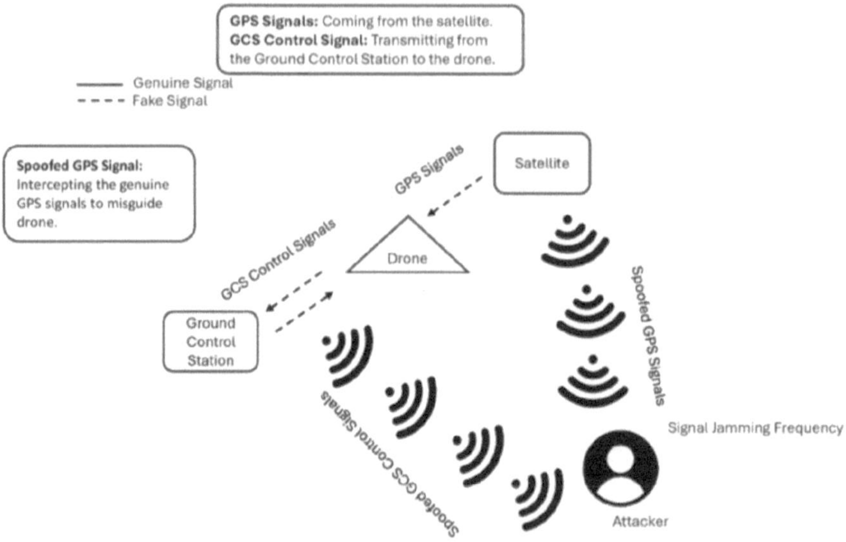

FIGURE 3.1 Jamming and spoofing attacks—illustration of signal spoofing and jamming affecting a drone's GPS and Ground Control Station (GCS) communications

A network mapper (Nmap) scanner was utilized in both assaults to gather accessible target devices.

Man-in-the-Middle Attack on AR Drone: To prevent man-in-the-middle attacks using WiFi Nano Pineapple, the writers kept an eye on the AR drone connection. When the Wi-Fi system's access points were examined, they resembled a pineapple's point of contact rather than a UAV. In order for an attacker to watch over the user's activities covertly, directives from the user were transmitted to the UAV. The application is susceptible to a man-in-the-middle attack, much like the UAV, because it tracks any other risky user-driven activity, such as visiting HTTP websites, rather than using the more secure Hypertext Transfer Protocol Secure (HTTPS) alternative.

Unauthorized AR Drone Root Control: File Transfer Protocol (FTP) and Telnet were two of the protocols used in this investigation. Deactivating the UAV involved using Telnet to gain access to the operating system's file system. Identification of the data was provided by the user ID, password, and many drone executables. By collecting and sending sensitive data to the attacker, an authorized user has taken control of the UAV, demonstrating a disruption in an attempt to reveal information.

AR Drone Packet Spoofing: The author tracks the vehicle using a spoofing node in the Python script for the AR drone, which imitates a UAV controller. First, the data packet is examined and identified, and the compatibility of the UAV controller is checked using the script. Subsequently, the cybercriminals employ an additional script to mimic the device's media access control (MAC) and IP addresses (Shafik, Matinkhah, & Shokoor, 2023).

3.2.5 Future Research Directions

Techniques for Adaptive and Lightweight Encryption: Creating resource-constrained UAVs with specific lightweight encryption techniques in mind. Investigating adaptive encryption methods as well, as this can dynamically modify parameters according to system circumstances.

Methods of Secure Wireless Communication and Countering Jamming Attacks: To offer dependable connection between UAVs and ground control stations, secure wireless communication protocols resistant to eavesdropping, jamming, and interception should be developed.

UAV Systems with Self-Healing and Resilience: Building robust UAVs that can recognize and neutralize cyberthreats on their own, maintaining business continuity in the face of cyberattacks.

Drone Threat Intelligence and Forensics: Gathering information on new cyberthreats that target UAV systems, evaluating it, and creating forensic analysis methods to look into and identify cyberattacks (Shafik, Matinkhah, & Shokoor, 2023).

3.2.6 General Attacks and Tools

Data Capturing and Forensics:

Attack: Capturing drone data can provide attackers with valuable information about the drone's operations, vulnerabilities, and potential targets. Forensics techniques may be employed to analyze captured data for investigative or intelligence purposes.

Tools/Mechanisms: Using serial connection, Extract DJI, Datcon, Prodiscover Basic

Location Tracing:

Attack: Tracing the location of drones can aid adversaries in tracking their movements, identifying operational patterns, and potentially targeting critical assets or infrastructure.

Tools/Mechanisms: Drone Monitoring Equipment, Acoustic Sensors, Radar

Man-in-the-Middle Attacks:

Attack: In order to eavesdrop on or modify data transmissions, man-in-the-middle (MITM) attacks entail intercepting and perhaps changing communication between drones and ground control stations (Sihag, Choudhary, Choudhary, & Dragoni, 2023).

Tools/Mechanisms: Wi-Fi attack, Remote AT Commands, WiFi Pineapple Nano, Raspberry Pi 3, Maldrone, SkyJack

3.2.7 Drones: Existing Threats on Data Security in Drones

Concerns with regard to drone security are raised by their extensive use across a range of sectors. Drones are utilized in transportation, medical, and military missions, but because of their intricate designs and dependence on numerous parts, they are vulnerable to both physical and cyberattacks.

Drone activities are immediately at risk due to the possibility of aircraft damage or crashes caused by physical attacks. Since drones primarily rely on communication channels to transmit sensitive data and monitor flight trajectories, cyber intrusions pose a more worrisome hazard. Cybercriminals are becoming more adept in breaking into drones over internet communication channels by employing advanced techniques.

Due to innate security flaws, commercial drones—particularly those written in C++—are especially susceptible. Strong security features are absent from many inexpensive drones, which raises the danger. Increasing user awareness is essential to reducing possible risks and fostering a safer drone environment.

The RQ-170 incident serves as a poignant example of drone vulnerabilities. During President Obama's tenure, Iranian forces captured and assumed control of the RQ-170 drone, prompting diplomatic tensions between nations. Such incidents underscore the urgency of addressing drone security vulnerabilities (Albalawi & Song, 2019).

Threat Mitigation:

Network Security: Strengthening Defenses Against UAV-Launched Attacks

i. Strengthening DHS networks is critical. UAS-launched cyberattacks are a genuine concern.
ii. Regularly deploying security updates, establishing firewalls, and limiting unwanted access are crucial procedures.
iii. Educating staff about IoT dangers helps lessen vulnerabilities.

Develop Defensive Concepts of Operations (CONOPS): Strengthening UAS Resilience

Randomized Deployment: UAS should not fly in predictable ways. Randomizing deployment sites confuses possible attackers.

Secondary Data Sources: UAS should include extra sensors (such as radar) to identify tampering or abnormalities.

Integrity Checks: UAS must do regular integrity checks throughout missions to guarantee that essential systems stay unmodified.

Minimizing Risk with Stringent Selection

Due Diligence: Policymakers and buying agencies must thoroughly assess UAS models.

Stringent Security Requirements: Our friends (UAS) must adhere to strict standards. Secure purchases minimize the risk of acquiring compromised or easily exploitable devices (Best et al., 2022).

3.2.8 Drones: Proposed Algorithms for Data Security

Secure Data Aggregation in WSN (Wireless Sensor Network): Data aggregation methods in WSNs aim to minimize the transmission of information between nodes by gathering data collected from various sensors and sent to the base station. These methods can be divided into structure-free and structure-based techniques. Structure-free methods face limitations related to nodes upstream and routing decisions. Structure-based techniques include flat network, cluster network, tree-based, and grid-based aggregation, each addressing specific energy consumption and efficiency concerns within the network.

Proposed Algorithm: The proposed event detection algorithm conducts a search for events based on specified requirements, notifying the base station upon detection. Subsequently, the data is transferred to the aggregator for aggregation before transmission to the base station. The tree-based aggregation algorithm, coupled with encryption methods, ensures data security and integrity during transmission.

Integration of seL4: The integration of the seL4 operating system aims to enhance data security in drones. Unlike traditional operating systems, seL4 employs a microkernel architecture, ensuring isolation between critical kernel processes and user services. Its bug-free nature and advanced security features, such as memory protection and clear communication protocols, make it suitable for UAV systems, addressing concerns related to cyber threats and data breaches.

3.2.9 Drones: How to Protect Them from Hacking Attacks

Drones are commonly utilized in sectors like deliveries and videography; however, they come with serious security threats from malevolent actors that want to steal data or interfere with operations. The integrity of drones can be jeopardized by flaws such as GPS spoofing and interception, which can target ground control stations (GCS), drones, and communication cables. It is advised to use intrusion detection systems, strong authentication procedures, and sophisticated cryptography techniques to improve security. Identification of criminals and comprehension of security breaches depend heavily on forensic techniques such as data carving and memory analysis. Adopting state-of-the-art forensic methods and preventive security measures will be crucial for maintaining resilience against new threats as drone technology develops (Semwal, Shikalgar, & Solanki, 2023).

3.3. ARTIFICIAL INTELLIGENCE

3.3.1 Introduction

Artificial intelligence (AI) mimics human cognitive abilities through algorithms and systems, revolutionizing machine interaction and data processing. Artificial Intelligence (AI) can examine data, spot patterns, and come to its own conclusions by using techniques like machine learning and deep learning. Applications of AI can be found in many different industries, including robotics, healthcare, finance, and self-driving cars. AI helps with algorithmic trading and fraud detection in finance, and it analyzes medical data to provide tailored treatment recommendations in the healthcare industry. Having a clear picture of the complicated terrain that artificial intelligence is associated with requires an understanding of both its technological features and its societal effects. We can exploit AI's potential to bring forth innovation, increase productivity, and raise living standards in a number of industries by using it safely and ethically.

3.3.2 Role of AI in Enabling UAV and Swarm Drones to Operate Effectively and Efficiently

Swarm drones can only function effectively and efficiently as a cohesive unit when artificial intelligence (AI) is used. AI enables swarm drones to sense, learn, plan, and act in dynamic and uncertain environments. It addresses a number of swarm drone topics, including formation of swarms, assignment of tasks, navigation, communication, decision-making, and adaptation. Swarm formation is the process of arranging drones into a particular pattern or shape, like a grid, a line, or a circle, to make a task easier to complete.

The process of allocating and dividing up responsibilities among the drones, such as finding, locating, and hitting targets, is known as task allocation. The practice of directing drones toward their intended destinations while averting collisions and impediments is known as navigation. The method by which the drones exchange signals and information, like their positions, velocities, or statuses, is referred to as communication. Making decisions involves deciding which course of action or approach is best for the drones given their objectives, preferences, and limitations. The process of modifying the drones' behavior or settings in response to their input, knowledge, or experience is referred to as adaptation (Semwal, Shikalgar, & Solanki, 2023).

3.3.3 State-of-the-Art Techniques and Algorithms

Swarm drone systems powered by AI use a variety of strategies, such as decentralized and centralized ones. While centralized approaches have advantages in global optimization, their scalability and resilience present issues. Centralized methods rely on a central controller for data collection and directive transmission. Decentralized approaches, on the other hand, take advantage of local drone interactions to function autonomously; however, they can run into coordination problems. Swarm drone systems may also use non-bio-inspired methods based on heuristics or mathematical models, or they may use bio-inspired techniques for optimization, such as modeling natural systems like bird flocks.

It is also important to distinguish between machine learning and non-machine learning techniques. Drones can learn from data by utilizing machine learning techniques like neural networks and reinforcement learning, experience, and user feedback to continuously improve. AI-driven swarm drone systems may effectively coordinate and optimize in a variety of scenarios while adjusting to changing conditions by strategically combining these technologies (Semwal, Shikalgar, & Solanki, 2023).

3.3.4 Key Directions and Research Opportunities for the Advancement of UAV and Swarm Drone Technology

Multi-objective Optimization Techniques: The answer to the question we are seeking is how is it possible at the present state of technological advancement in this aspect to enable swarm drones to optimize multiple objectives simultaneously, such as minimizing the cost, maximizing the performance, and satisfying the constraints, while balancing the trade-offs and conflicts among them.

Human-Swarm Interaction: Another interesting one would be the modus operandi in bringing to life, a technique by which swarm drones can engage with humans in a natural, intuitive, and user-friendly way, such as using voice, gesture, or haptic commands, while respecting the human's preferences, expectations, and emotions.

Swarm Drones in Industrial Applications: The swarm drones could potentially possess the caliber to perform various tasks in industrial settings, such as manufacturing, logistics, inspection, and maintenance, while integrating with the existing systems, processes, and standards.

Bio-inspired Algorithms and Swarm Intelligence: A very captivating intersection would be the swarm drones learning from and being able to emulate the behavior or mechanisms of natural systems, such as flocks of birds, schools of fish, and colonies of ants, to solve complex optimization and coordination problems, while enhancing their adaptability, diversity, and creativity.

Explainability and Interpretability in AI-Driven Swarm Systems: Swarm drones could become the best solutions out there to explain and justify their own actions, decisions, and strategies, in a transparent, understandable, and trustworthy way, while providing insights, feedback, and recommendations to the users, stakeholders, and regulators (Semwal, Shikalgar, & Solanki, 2023).

3.3.5 Advancements in UAV Path Planning Technology

Path planning is crucial to UAV operations in order to generate waypoints by leveraging environmental and UAV attitude data. While meta-heuristic algorithms like PIO, FOA, GWO, and PSO use real-time data, traditional algorithms like RMA, A*, and APF rely on preloaded environmental information. Finding the best route between the start and end locations is the main goal of UAV path planning. Area coverage, dynamic route planning, ideal road strength, and 3D path planning are among the components that are the focus of research on UAV cluster path planning. The effectiveness and efficiency of UAV operations are strongly impacted by each of these criteria. Efficient route planning guarantees safe and easy navigation for UAVs, which enhances their overall performance in a range of tasks and uses. Additional techniques for real-time route optimization are provided by algorithms such as PIO, FOA, GWO, and PSO, which improve the flexibility and responsiveness of UAV operations.

3D Path Planning: Finding practical and straightforward pathways between places is a difficulty for multi-UAV systems using 3D path planning. An enhanced Gray Wolf Optimization (GWO) technique was employed by Dewangan R K and colleagues to produce feasible flight paths that circumvent obstructions. A multi-trajectory planning approach based on a 3D Probabilistic Road Map (PRM) allows UAV swarms to explore various sites efficiently in urban emergencies. Using a multi-swarm Fruit Fly Optimization Algorithm (MSFOA), a multi-drone collaborative path planning method is used to address path planning in hazard areas, particularly on rough terrain. The Fruit Fly Optimization Algorithm (FOA) carries out local optimization for obstacle avoidance, whereas the Pigeon Inspired Optimization (PIO) algorithm improves starting pathways for 3D oilfield detection. When crossing static rough terrain, factors like path length, height, and adjustment angle are critical. To increase the effectiveness and efficiency of multi-UAV systems, enhanced particle swarm optimization (PSO) techniques are used to consider path planning as a three-objective optimization problem.

Dynamic Path Planning: For UAVs operating in complicated surroundings, dynamic path planning is crucial, necessitating constant obstacle avoidance and quick reaction to potential dangers. Candidate paths are generated by an algorithm based on the cubic spline second-order continuity principle, and the optimal path is chosen using a total cost function. Using complementary sensors to build an adaptive route takes into account unknown situations that may change. The acyclic problem of the Wall-Following Method (WFM) and the local minimum problem of the Artificial Potential Field (APF) approach are solved by techniques such as these. A framework for tracking direction estimation with lower computation and input, based on Support Vector Machines, facilitates track recognition and automatic scene understanding in difficult outdoor contexts. The effectiveness and efficiency of multi-UAV systems are greatly increased by preventing path conflicts in response to inconsistent communication status data from numerous drones.

Optimal Path Planning: The growing importance of Unmanned Aerial Vehicles (UAVs) and the Internet of Things (IoT) has made optimal path planning technologies crucial. Because UAVs have a limited amount of battery life, mission execution requires careful planning of the best routes. Research on fixed-wing UAV-assisted mobile crowd sensing (MCS) systems has been concentrated in this sector. An energy efficiency approach is applied to the study of joint path planning and work allocation problems. The original NP-hard joint optimization issue is reformulated as a bilateral two-stage matching problem with positive matching performance, overall profit, and energy consumption.

Efficient data collection with minimal energy consumption is essential to guarantee terminal compliance and optimize drone and sensor power usage. A coupled distributed planning method that combines task assignment and trajectory generation is used to calculate the ideal flight trajectory when a collection of heterogeneous fixed-wing UAVs visit various targets and complete continuous tasks. Cooperative task planning issue is recreated under some assumptions and relaxed Dubins path. This technique offers potential for real-world use and greatly increases the system's running rate. This all-encompassing method of optimal path planning technology greatly improves multi-UAV systems' efficacy and efficiency.

Coverage Path Planning: For UAVs, Coverage Path Planning (CPP) entails creating trajectories that cover the entire area while making use of spatial data and a Dual-Depth Q-Network (DDQN) to make control decisions. An enhanced Binary Logarithmic Linear Learning (BLLL) algorithm is used to solve the coverage search problem, allowing numerous drones to be coordinated and reducing trajectory oscillation through reciprocal techniques. Hybrid algorithms for enhanced detecting capability and clustering techniques for tracking moving targets address issues including coverage, location, and orientation.

In areas full of obstacles, a gradient algorithm (GA) with improved depth deterministic algorithms guides UAVs to follow ground objects, penalizing smooth trajectories and deploying several drones for increased detection performance. By simulating environmental states, long and short memory networks enhance the effectiveness and efficiency of multi-UAV systems (Zhou, Rao, & Wang, 2020).

3.3.6 AI Use Case

AI-Based UAV in Traffic Monitoring: There are various difficulties with using AI-powered UAVs for traffic surveillance. These include managing a sizable video feed and real-time image capture, overcoming challenges with data collecting in city settings, identifying irregularities and setting off alarms, and connecting with the entire traffic system. Predictive traffic management systems, fleets of autonomous UAVs, Swarm Intelligence implemented in UAVs, efficient Edge-Computing integration, and the development of more sophisticated AI for managing large amounts of data are all included.

AI in UAV for Object Detection: Object detection using AI in UAVs faces challenges such as detecting small and moving objects, dealing with the complexity due to non-annotated data, and adapting to diverse environments. The future scope in this area involves creating domain-specific transfer learning, privacy-aware object detection techniques, Edge AI and onboard processing, and enhancing generalization and transferability.

AI-Enabled Agriculture Surveillance UAV System: AI-enabled UAVs in agriculture encounter difficulties with data administration and storage, classifying crops and pests, identifying them among many classes, integrating with current farming systems, and resolving problems with battery life and long-range flight endurance. Future capabilities include automated AI-based agricultural data analysis, hyperspectral imaging, integration with IoT devices for enhanced analysis, autonomous navigation, and dynamic planning.

Rescue Operation Surveillance with UAV Embedded AI: AI-enabled UAV rescue missions must overcome obstacles such as traversing hazardous terrain, controlling payload and endurance, and maintaining situational awareness in real time. Future developments in this field could include strong design and redundancy development, AI-assisted medical support, and AI integration for disaster prediction (Pal, Shovon, Mridha, & Shin, 2023).

3.3.6.1 Case Study 1: Swarm Drones and AI for Emergency Response: Case Study of Northern California Wildfires (Nov. 2018)

In late 2018, a large-scale drone emergency response aided Northern California post devastating wildfires. DJI, the leading drone manufacturer, partnered with local authorities and private firms to map 17,000 acres in two days. The aim is to create damage maps for insurance, aid planning, recovery, and SAR missions. Challenges included delayed site access, but 16 UAV teams conducted 518 flights, capturing 70,000+ images. Processing 500GB data proved challenging, resolved by transferring it to Drone Deploy HQ, resulting in 75 maps overnight.

AI was pivotal, especially in detecting damaged structures. Unleash Live's AI method enabled insights extraction on structural and vehicle damage as shown in Figure 3.2. Trained on high-resolution datasets and past wildfire imagery, the AI model achieved 80–90% accuracy. The AI algorithm swiftly processed drone imagery, automating structure, and vehicle detection. This case study underscores drones and AI's efficacy in emergency response, facilitating rapid assessment and decision-making during crises, thus enhancing disaster management operations (Oren & Verity, 2020).

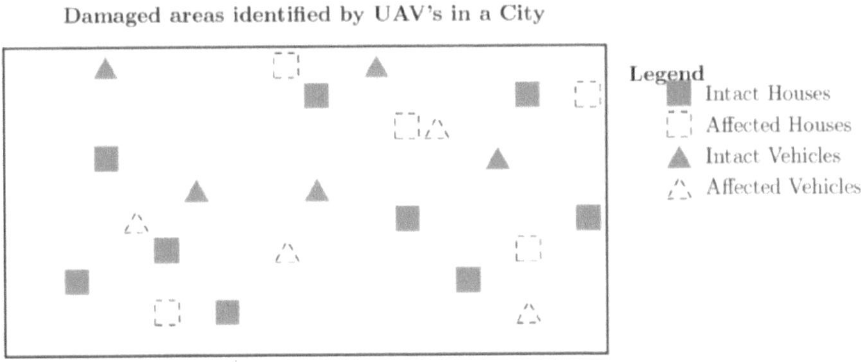

FIGURE 3.2 UAV identifying damaged areas—aerial imagery analysis for post-disaster damage assessment that separates the damaged and intact structures

3.3.6.2 Case Study 2: Swarm Drones and AI for Emergency Response: Case Study of Cyclone Idai in Malawi (Mar. 2019)

In March 2019, Cyclone Idai wreaked havoc in Malawi, Zimbabwe, and Mozambique, affecting over 700,000 people. As part of the emergency response, UNICEF deployed drones in Malawi to assess the damage caused by floods and heavy rains in Nsanje District. The three-person UNICEF team mapped the worst-affected areas in five days by using the drone corridor set up at Kasungu Aerodrome, giving crucial aerial data to support disaster response activities. They rapidly determined the degree of flooding and damage across eight square kilometers using Pix4D's high-resolution imagery and mapping software.

The team used algorithms for machine learning and artificial intelligence (AI) created by Globhe. Quick decision-making and suggestions for community planning were made possible by these algorithms' automatic identification of residences, flooded regions, and other vital data. Despite the success of the AI analysis, challenges emerged, particularly in detecting grass-thatched roofs among the affected structures. However, the team's innovative approach, combined with drones and AI technology, enabled ground teams to identify at-risk populations, assess infrastructure damage, and plan recovery efforts efficiently. This case study highlights the effective integration of drones, AI, and machine learning in emergency response efforts, demonstrating their potential to enhance disaster preparedness, response, and recovery in vulnerable communities (Oren & Verity, 2020).

3.3.7 Drone Obstacle Avoidance by Reinforcement Learning

Introduction to Reinforcement Learning in Artificial Intelligence: A state-of-the-art method in artificial intelligence (AI) is deep reinforcement learning (DRL), which combines the ideas of reinforcement learning and deep learning. By interacting with complicated surroundings, it enables AI entities to learn optimal behaviors, which is similar to the process of trial-and-error learning seen in humans and animals. Deep neural networks (DRL) allow agents to efficiently process high-dimensional input, leading to notable advancements in a range of AI applications, from gaming proficiency to autonomous system control.

RL algorithms can be broadly categorized into three main types:

Value-Based Methods: These algorithms select activities that maximize the predicted cumulative benefit by estimating the value of doing particular actions in various phases.

Policy-Based Methods: These algorithms do not explicitly estimate the value of actions; instead, they learn a policy, or a mapping from states to actions, directly.

Actor-Critic Methods: These algorithms maintain distinct networks to estimate the policy and the value function, combining elements of value-based and policy-based approaches.

UAV Navigation Based on Reinforcement Learning: In order to maximize decision-making, reinforcement learning (RL) places a strong emphasis on interaction with the environment and evaluative feedback signals. It uses a trial-and-error approach akin to that of human and animal learning. Deep reinforcement learning (DRL), in particular, is a type of reinforcement learning (RL) that shows promise in addressing challenging optimization issues like UAV navigation.

Value function-based and policy-based reinforcement learning are two subtypes of RL. Although Q-learning is a popular technique in value function-based reinforcement learning, its applicability to dynamic situations is limited due to its discrete state and action spaces. Deep Q Network (DQN) is one of the innovative algorithms used in DRL, which integrates deep learning with reinforcement learning to address this. However, path planning quality can still be enhanced due to DQN's discrete action space.

Policy-based RL algorithms such as Distributed Proximal Policy Optimization (DPPO) and Deep Deterministic Policy Gradient (DDPG) have been developed to handle continuous state and action spaces. These algorithms, while useful for UAV path planning, suffer from time-consuming training and convergence problems in complicated environments. While visual perception is essential for UAV navigation, vision-based DRL techniques require a lot of resources to train. Certain methods reduce the complexity of images to compress data while maintaining important aspects. To help in obstacle avoidance, models based on variational autoencoders (VAEs) are frequently employed to extract low-dimensional feature information from pictures.

When VAE and policy-based DRL algorithms are combined, a lightweight network structure is provided, which speeds up convergence and maximizes resource use for UAV obstacle avoidance. The capacity of UAVs to efficiently navigate complicated settings is improved by this integration (Xue & Gonsalves, 2021).

SAC (Soft-Actor-Critic): With an emphasis on continuous control, the Soft-Actor-Critic (SAC) framework is a deep reinforcement learning (DRL) technique designed for applications such as drone navigation. Comparing SAC to other DRL techniques, the combination of maximum-entropy reinforcement learning and actor-critic architecture improves robustness and exploration capabilities. By maximizing entropy and dynamically modifying entropy weight during training, it promotes exploration by concurrently optimizing the policy's action distribution entropy and expected accumulated rewards. SAC interacts with the environment by enabling policy optimization based on quality evaluations of state-action pairings by storing data about states, actions, rewards, and subsequent states in a buffer (Xue & Gonsalves, 2021).

VAE (Variational Auto-Encoder): Variational Auto-Encoders (VAEs) are used in machine learning for image generation, dimensionality reduction, and representation learning. In the first training phase, VAEs work with SAC-like reinforcement learning algorithms to produce latent codes that allow VAEs to create compressed representations of input data, like depth maps. These latent codes are essential for applications like drone navigation because they efficiently capture important features and lower dimensionality. Through the provision of appropriate state representations, VAEs improve the performance and efficacy of reinforcement learning algorithms such as SAC, guaranteeing quick convergence while preserving pertinent state information (Xue & Gonsalves, 2021).

Deep Learning-Based Object Detection in UAV: The development of object detection algorithms is divided into two stages: deep learning techniques and conventional methods. Conventional techniques with constraints in computational complexity and feature representation include sliding window and artificial feature extraction. This method made extensive use of the Viola-Jones detector and the Histogram of Oriented Gradients (HOG). Convolutional neural networks (CNNs), in particular, have revolutionized object detection through deep learning by providing high speed, resilience, and automatic high-level information extraction.

There are two main approaches for object detection using deep learning: one-stage and two-stage methods. From Figure 3.3, two-stage detectors, such as Faster R-CNN, work by first suggesting regions, followed by classification and regression, to address boundary detection and imprecise localization problems (Tang, Ni, Zhao, Gu, & Cao, 2023).

In contrast, one-stage detectors from Figure 3.4, such as SSD (Single Shot MultiBox Detector) and YOLO (You Only Look Once), anticipate item location and category immediately from the image, doing away with the requirement for proposal regions and speeding up processing. They might have trouble precisely localizing and identifying small objects, as well as collecting finer details, which could result in inconsistent performance.

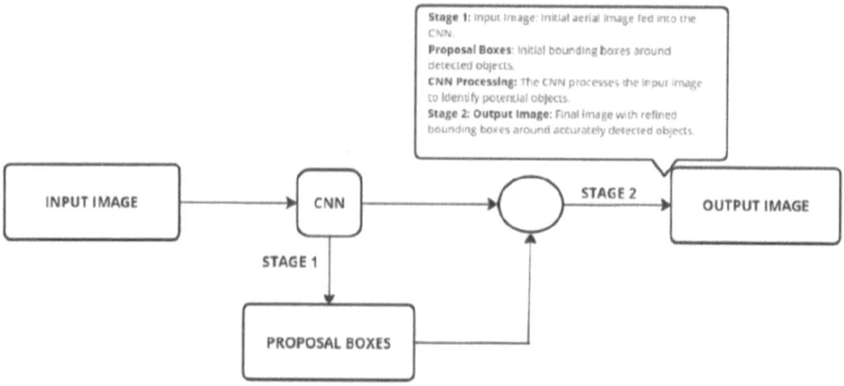

FIGURE 3.3 Two-stage object detection framework—two-stage object detection process using convolutional neural networks (CNN). The input image is processed to generate proposal boxes, which are then refined in the second stage to produce the final output image with detected objects.

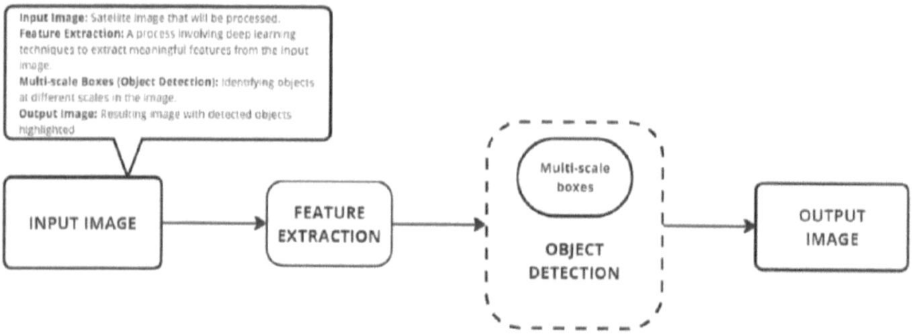

FIGURE 3.4 One-stage object detection framework: A deep learning-based object detection process for satellite imagery. The input image undergoes feature extraction, followed by object detection using multi-scale boxes, resulting in an output image with detected objects highlighted.

In order to address these unique difficulties encountered in UAV applications including geological environment surveys, traffic monitoring, and precision agriculture, techniques like CPNet, NQSA, SGMFNet, and RSOD have been proposed to increase detection accuracy, speed, and robustness (Tang, Ni, Zhao, Gu, & Cao, 2023).

3.4 DATA SCIENCE

3.4.1 Introduction

To examine both organized and unstructured data, data science combines domain-specific knowledge, computer science, statistics, and mathematics. Large-scale data processing, analysis, and gathering utilizing cutting-edge computer methods are all part of it. Data collection from many sources, exploration, visualization, statistical analysis, and machine learning are important elements. Big data technologies, such as Apache Hadoop and Spark, are used to efficiently handle enormous datasets. Decision-making in sectors including manufacturing, banking, healthcare, e-commerce, and telecommunications is fueled by data science, which also produces insights. Predictive maintenance and personalized recommendations are just two of its uses. Because data-driven decision-making is becoming increasingly important, skilled data scientists are in high demand.

3.4.2 Fundamentals of Swarm Intelligence Algorithms

SI algorithms work on the concepts of simplicity, scalability, robustness, and adaptability. These algorithms begin with the initialization of a population of candidate solutions, which then collaborate and evolve across iterations to achieve optimal results. Various SI algorithms have been developed over time, drawing inspiration from a variety of natural phenomena such as birds, insects, animals, and primitive organisms. These algorithms use a variety of operations and techniques to successfully explore solution spaces (Chopra & Arora, 2022).

3.4.3 Real-World Applications of SI Algorithms

Beyond theoretical applications, SI algorithms show extraordinary efficacy in real-world circumstances across several areas. SI algorithms help discover communities in social networks by revealing implicit community patterns inside networks. Scheduling and routing issues, which are common in many sectors, use SI-based techniques for optimization. SI algorithms also have applications in bioinformatics, resource allocation, Internet of Things (IoT) systems, and anomaly detection activities, demonstrating their variety and adaptability.

3.4.4 Challenges, Opportunities, and Limitations

Despite their effectiveness, SI algorithms confront several obstacles and limits. One significant obstacle is the universality of the principles controlling SI algorithms across diverse problem areas. Furthermore, developing effective SI algorithms requires a thorough grasp of their core components and operations. While SI algorithms provide promising answers for high-dimensional and dynamic data, further study is needed to improve their ability to properly handle such data.

3.4.5 Future Directions

The future of Swarm Intelligence algorithms in Data Science has enormous promise for breakthroughs and discoveries. Researchers are ready to tackle current issues and investigate new potential for developing more efficient and flexible SI algorithms. Integrating SI techniques with machine learning frameworks for autonomous optimization and model construction is an exciting prospect. Improving the learning modules and assessment mechanisms inside SI algorithms would help to increase their acceptance and efficacy in handling complicated data science problems (Chopra & Arora, 2022).

3.4.6 Introduction to Data Science in UAV Agriculture

In the field of UAV agriculture, data science serves as the foundation for making educated decisions. As drones capture massive volumes of agricultural data, the role of data science becomes critical in deriving useful insights. These findings contribute to current agriculture methods' efficiency, productivity, and sustainability. With sophisticated analytics, UAV operators may successfully optimize resource allocation, monitor crop health, and prevent hazards. The use of data science with UAV agriculture represents a substantial move toward precision farming and data-driven management approaches. By leveraging data science, players in UAV agriculture may create new potential for innovation and growth.

3.4.7 Operations of Standard Agricultural Data (Drone Network)

Standard big data operations in UAV agriculture entail a methodical approach to data collection, processing, and analysis. Communication service providers can enhance operational efficiency and resource allocation by studying network data. Data-driven operations help stakeholders make educated decisions about everything from crop health to market trends. Integrating data analytics into UAV agriculture operations boosts productivity, lowers costs, and promotes sustainable practices. By using standard agricultural data, stakeholders can open up new avenues for innovation and growth in the UAV agriculture sector (Meivel & Maheswari, 2020).

3.4.8 Algorithms Relevant to the Analysis and Management of Agricultural Data Within Drone Networks

Random Forest Algorithm: Figure 3.5 shows that this approach is used for classification and regression problems. Class weight estimates may be used to measure mappers' training sets in agricultural data analysis. During training, the method creates numerous outputs and decision trees, the class that represents the mean prediction of each individual tree (regression) or the mode of the classes (classification). The text states that the random forest technique is transformed using Random Forest CS without leaves weights and classes, implying a tailored implementation for certain use cases (Meivel & Maheswari, 2020).

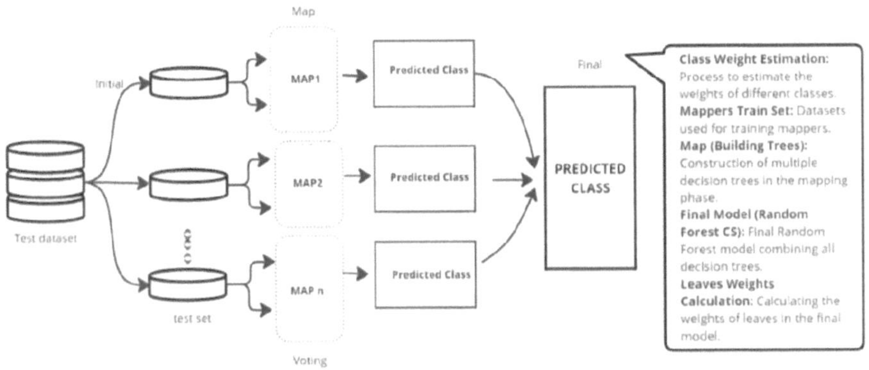

FIGURE 3.5 Random Forest Algorithm: train a Random Forest Classifier using Map Reduce—estimate class weights, split the dataset, build decision trees in parallel mappers, and combine them for the final model with optimized splits, leaf classes, and weights.

K-Nearest Neighbors (KNN) Algorithm: KNN is a non-parametric approach to both supervised and unsupervised learning. The KNN method is mentioned in the material, as well as the use of MAP partitioning and key reduction. This shows that KNN is utilized in agricultural drone networks to analyze geographical data or perform clustering tasks.

MapReduce: MapReduce is a programming idea and implementation that uses a distributed, parallel algorithm on a cluster to process and create massive datasets. There are two steps involved: a map step that breaks down the incoming data into more manageable subproblems and a reduction step which aggregates the results of the map stage. MapReduce helps agricultural drone networks efficiently combine and process large amounts of data (Meivel & Maheswari, 2020).

Oversampling and Undersampling: These techniques solve class imbalance in datasets and are highly beneficial for classification tasks. Figure 3.7 shows that while oversampling involves raising the number of cases in the minority class, undersampling requires a decrease in the number of instances in the majority class. The data includes both oversampling and undersampling rates, which are employed in agricultural drone networks for data pre-processing and balancing (Meivel & Maheswari, 2020).

3.5 CONCLUSION

A revolutionary horizon with significant consequences is presented by the convergence of cybersecurity, data science, and artificial intelligence (AI) in swarm drones and autonomous systems. Although these technologies make it possible to collaborate and operate effectively in dynamic contexts, they also present serious cybersecurity risks, such as privacy violations and hacking vulnerabilities. Ensuring the appropriate deployment of AI-powered technologies requires strong cybersecurity

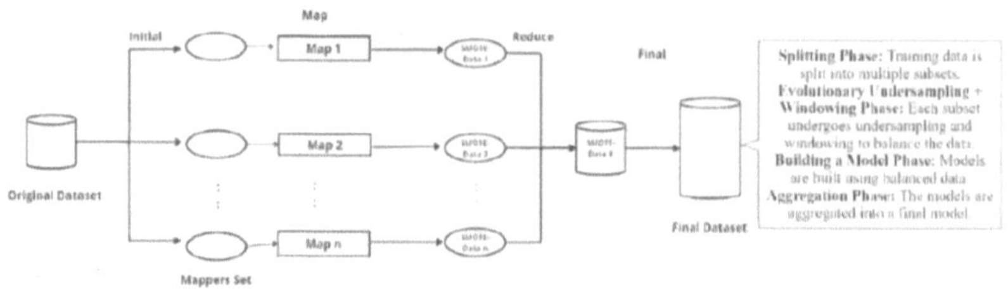

FIGURE 3.6 Oversampling—SMOTE with MapReduce divides the original dataset, oversamples using SMOTE, combines the processed data, and produces a final balanced dataset.

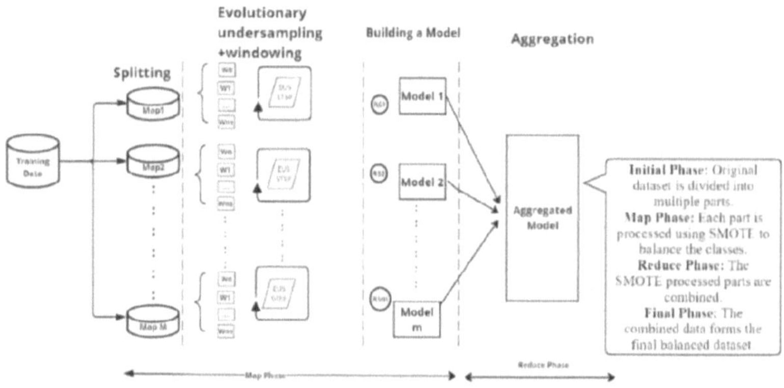

FIGURE 3.7 Undersampling—evolutionary undersampling framework splits the training data into subsets, applies undersampling and windowing, builds models, and aggregates them.

protocols. To fully solve these issues and realize the promise of autonomous systems and swarm drones, cooperation between researchers, legislators, and industry stakeholders is vital. Together, we can handle the challenges posed by cybersecurity, take advantage of the caliber of AI and data science, and thus, mitigate risks to pave the path for a safe and creative future for aerial operations.

REFERENCES

Albalawi, M., & Song, H. (2019, April). Data security and privacy issues in swarms of drones. In *2019 Integrated Communications, Navigation and Surveillance Conference (ICNS)* (pp. 1–11). IEEE.

Best, K. L., Schmid, J., Tierney, S., Awan, J., Beyene, N. M., Holliday, M. A., Khan, R., & Lee, K. (2022). *How to analyze the cyber threat from drones.* RAND Corporation. https://doi.org/10.7249/RRA1376-1

Chopra, D., & Arora, P. (2022). Swarm intelligence in data science: Challenges, opportunities, and applications. *Procedia Computer Science, 215*, 104–111.

Meivel, S., & Maheswari, S. (2020). Standard agricultural drone data analytics using KNN algorithm. *Test Engineering and Management, 82*, 206–215.

Oren, C., & Verity, A. (2020). *Artificial intelligence (AI) applied to unmanned aerial vehicles (UAVs) and its impact on humanitarian action.* Digital Humanitarian Network, Lille, France.

Pal, O. K., Shovon, M. S. H., Mridha, M. F., & Shin, J. (2023). A comprehensive review of AI-enabled unmanned aerial vehicles: Trends, vision, and challenges. *arXiv preprint* arXiv:2310.16360.

Semwal, A., Shikalgar, S., & Solanki, R. (2023). The use of artificial intelligence in swarm drones. *International Journal for Research in Applied Science & Engineering Technology, 11*(7), 1052–1057.

Shafik, W., Matinkhah, S. M., & Shokoor, F. (2023). Cybersecurity in unmanned aerial vehicles: A review. *International Journal on Smart Sensing and Intelligent Systems, 16*(1).

Sihag, V., Choudhary, G., Choudhary, P., & Dragoni, N. (2023). Cyber4Drone: A systematic review of cybersecurity and forensics in next-generation drones. *Drones, 7*(7), 430.

Tang, G., Ni, J., Zhao, Y., Gu, Y., & Cao, W. (2023). A survey of object detection for UAVs based on deep learning. *Remote Sensing, 16*(1), 149.

Xue, Z., & Gonsalves, T. (2021). Vision-based drone obstacle avoidance by deep reinforcement learning. *AI, 2*(3), 366–380.

Zhou, Y., Rao, B., & Wang, W. (2020). UAV swarm intelligence: Recent advances and future trends. *IEEE Access, 8*, 183856–183878.

4 Drones in Delivery and Logistics Transforming Transportation Systems

Aditya Srivastav and Shishir Kumar Shandilya

4.1 INTRODUCTION

4.1.1 Overview of Drones in Delivery and Logistics

Drones have emerged as a disruptive force in the delivery and logistics sectors, offering new solutions to meet modern supply chain demands. Unlike traditional delivery methods, drones can bypass ground traffic, delivering goods quickly and directly to customers, even in densely populated urban areas or remote locations with limited infrastructure. The ability of drones to perform last-mile deliveries—often the most challenging and costly part of logistics—has captured the interest of companies looking to improve efficiency and customer satisfaction.

In real-world applications, drone integration into logistics networks has been pioneered by corporations such as Amazon. Amazon's Prime Air program, for example, aims to use autonomous drones to deliver packages within 30 minutes of ordering. This initiative not only showcases the speed and efficiency of drone delivery but also highlights the growing consumer demand for faster shipping options. Another notable example is Zipline, which uses drones to deliver critical medical supplies, such as blood and vaccines, to remote areas in Rwanda and Ghana. This has proven to be a life-saving application, particularly in regions with poor road infrastructure or during emergencies where timely delivery is crucial (Enarbia & Kyamakya, 2021).

4.1.2 Evolution of Drone Technology

Drone technology has rapidly evolved from its origins in military applications to a versatile tool used in various civilian industries, including logistics. Early drones were primarily used for reconnaissance and surveillance, but advancements in technology have expanded their capabilities significantly. Key developments include:

- GPS Navigation and Autonomy: Modern drones have sophisticated GPS systems installed and AI-based navigation technologies that allow for autonomous flight and precise route optimization. These technologies

DOI: 10.1201/9781003481584-4

enable drones to operate with minimal human intervention, making them highly efficient for delivery tasks (Raj & Sah, 2019).

- Battery and Power Systems: The development of high-capacity lithium-ion batteries has significantly expanded the flight range and payload capacity of drones. To further improve drone performance and shorten charge periods, researchers are still investigating novel battery technologies including solid-state batteries (Peng et al., 2019).
- Artificial Intelligence (AI) and Machine Learning: AI plays a crucial role in enabling autonomous drone operations. Machine learning algorithms help drones optimize flight paths, avoid obstacles, and even make real-time decisions based on environmental data. These advancements have made drones more reliable and capable of handling complex delivery tasks.
- Materials Science: The utilization of sturdy and lightweight materials, such as carbon fiber, has improved the structural integrity and payload capacity of drones. These materials help drones to be more energy-efficient and capable of carrying heavier loads over longer distances (Chiang et al., 2019).

UPS Flight Forward is a prominent example of how these technological advancements have been applied in logistics. The company has taken advantage of the utility of autonomous drones to deliver medical supplies between healthcare facilities, highlighting the potential for drones to enhance the efficiency and reliability of time-sensitive deliveries (Law et al. 2023).

4.1.3 Current Trends and Market Dynamics

Several key trends and market dynamics influence the adoption of drones in delivery and logistics:

- E-commerce Growth: The explosive growth of e-commerce has fueled the demand for fast and efficient delivery services. Drones are increasingly seen as a solution to meet these demands, particularly for last-mile deliveries, where speed and flexibility are crucial. Companies like Amazon have been at the forefront of this trend, experimenting with drone delivery to enhance customer satisfaction and streamline logistics operations (Rejeb et al., 2021).
- Sustainability and Environmental Impact: Drones offer a more sustainable alternative to traditional delivery vehicles, especially in urban areas. Their electric-powered engines result in lower carbon emissions, aligning with the growing emphasis on sustainability in corporate strategies. Research shows that optimally routing drones for deliveries can significantly reduce the carbon footprint of logistics operations (Rejeb et al., 2021).
- Regulatory Developments: The regulatory landscape is evolving to accommodate the growing use of drones in commercial delivery. To guarantee the safe integration of drones into national airspace, regulatory agencies like the FAA in the US and the EASA in Europe are creating standards. These

regulations are critical in determining the scope and scale of drone deployment in logistics (Stolaroff et al., 2018)

- Public Perception and Acceptance: Public perception of drones is gradually shifting as people become more accustomed to seeing them in everyday life. However, concerns about privacy, safety, and noise pollution remain. Companies involved in drone logistics are investing in public education and community engagement to address these concerns and build trust. Demonstrating the benefits of drone delivery, such as reduced traffic congestion and faster emergency response times, is key to gaining public acceptance.

Walmart and Google's Wing are examples of companies exploring drone delivery for retail and grocery items, seeking to enhance customer convenience and operational efficiency. These companies are conducting pilot programs and exploring partnerships with regulatory agencies to expand their drone delivery services. In rural areas, companies like Zipline are using drones to overcome logistical challenges posed by poor infrastructure, ensuring that critical supplies reach those in need more efficiently (Nurgaliev et al., 2023).

Technological developments, shifting consumer demands, and changing legal frameworks are driving the incorporation of drones into delivery and logistics. As companies continue to innovate and adapt to these trends, drones are set to become an increasingly integral part of the global logistics network.

4.2 ADVANTAGES OF DRONES IN LOGISTICS

4.2.1 Speed and Efficiency

Drones offer significant advantages in speed and efficiency, especially in last-mile delivery. Drones can avoid these problems by flying straight to the delivery location, in contrast to traditional delivery methods that are frequently hampered by traffic congestion, road infrastructure, and other delays. Faster delivery times are made possible by this capacity, which is especially helpful in cities with heavy traffic. Drones are capable of delivering goods within minutes, a feat that is unattainable with conventional ground transportation. This rapid delivery capability is particularly appealing to e-commerce companies and consumers who prioritize fast service. Studies have shown that optimizing drone routes can enhance the efficiency of deliveries, reducing both time and operational costs (Das et al., 2021)—which is comparatively less expensive and environmentally friendly than gasoline or diesel. Lower fuel consumption not only saves costs but also corresponds to sustainable goals. In the case of last-mile delivery, where these aircraft can complete traditional deliveries more quickly or replace them altogether, drones could also help keep expenses down in the area of logistics via cost-effective operations around drones proved by one study (Raj & Sah, 2019). Drones also help lower the carbon emissions related to logistics operations, reducing their carbon footprint and consequent savings in carbon credits collections and other kinds of environmental rewards (Chiang et al., 2019).

4.2.2 Environmental Benefits and Sustainability

Drones' beneficial effects on the environment are among their most important logistics benefits. When compared to conventional delivery trucks, drones—especially those that run on rechargeable batteries—produce noticeably fewer greenhouse gas emissions. This reduction in emissions contributes to the sustainability of logistics operations, making drones a greener alternative for parcel delivery. Research has shown that drones can reduce the environmental footprint of delivery services by minimizing energy consumption and decreasing the production of pollutants. For instance, a study comparing drone and motorcycle deliveries found that drones produce substantially lower greenhouse gas emissions and particulate matter, especially in rural areas where the environmental benefits are magnified due to the longer distances between delivery points (Park et al., 2018). Integrating drones into logistics can help companies meet stringent environmental regulations and consumer demands for more sustainable practices.

4.2.3 Accessibility to Remote Areas

Drones are particularly effective in reaching remote or hard-to-access areas where traditional delivery vehicles may struggle due to poor infrastructure or natural barriers. This makes them invaluable for delivering essential goods, such as medical supplies, to rural or disaster-stricken regions where road access is limited or non-existent. Drones have been effectively employed by businesses such as Zipline to provide medical supplies and blood to isolated areas of Rwanda, significantly improving healthcare delivery in these areas (Gunaratne et al., 2022). Drones' ability to operate independently of road networks also makes them a reliable solution for emergency response, where timely delivery of supplies can be life-saving.

Drones offer numerous advantages in logistics, from enhancing delivery speed and efficiency to providing cost savings and promoting environmental sustainability. Their ability to access remote areas further underscores their potential as a transformative technology in the logistics sector.

4.3 TYPES OF DELIVERY DRONES

4.3.1 Fixed-Wing Drones

Fixed-wing drones are a staple in aerial operations that demand long-range capabilities and high-speed performance. These drones are particularly energy-efficient, enabling extended flight durations that are essential for covering vast areas or executing long-distance deliveries. This makes them ideal for applications such as agricultural monitoring, environmental surveys, and logistics over remote regions. However, their design necessitates specific launch and recovery systems, such as runways or catapult systems, which can limit their deployment in confined or rugged environments. The fixed-wing configuration is less maneuverable compared to other drone types, making it less suitable for operations that require precision in tight spaces. The complexity of their operation and maintenance also tends to be higher, requiring specialized knowledge and infrastructure.

4.3.1.1 Use Cases and Applications in Logistics

Fixed-wing drones are particularly useful in logistics for tasks requiring long range and endurance. They are commonly used in situations where large areas need to be covered, like when medical supplies are delivered to isolated or underprivileged areas. For example, Zipline's fixed-wing drones have played a crucial role in providing medical supplies and blood to isolated regions of Rwanda and Ghana, significantly reducing delivery times and improving healthcare outcomes (Law et al., 2023).

Another application is in the e-commerce sector, where fixed-wing drones can facilitate long-distance deliveries more efficiently than traditional methods. Companies like Amazon have explored using fixed-wing drones for their Prime Air service, aiming to deliver packages to customers within 30 minutes (Benarbia & Kyamakya, 2021).

In addition to medical and e-commerce deliveries, fixed-wing drones are also employed in agricultural logistics, particularly in large-scale farming operations where they can monitor crop health, manage irrigation, and apply fertilizers or pesticides efficiently over vast fields (Godbole et al., 2019).

4.3.2 Rotary-Wing Drones (Quadcopters, Hexacopters, etc.)

The capabilities of rotary-wing UAVs (rotary-wing drones or quadcopters/helicopters) to Vertical Take-Off and Landing (VTOL) are of high value. This also allows them to operate in nearly any environment, from dense cities to remote, rough-and-tumble terrain, without requiring runways. Their outstanding performance in close quarters, maneuverability, and stability, the latter two of which facilitate smooth navigation in very tight spaces, are hard to beat for last mile delivery work, inspections, and emergency services. According to the design principles, helicopter type UAVs have certain limitations typically associated with shorter flight distances and lower maximum velocities as compared to fixed-wing drones. They also expend more energy and have shorter battery lives and lower mission persistence. And they're typically limited in the amount of cargo they can carry, causing limitations for more weighty gear or goods.

4.3.2.1 Use Cases and Applications in Logistics

Helicopter drones are especially beneficial in urban areas since they require less space to take off and land. They are meant for so-called last-mile delivery, as they are able to use sidewalks to meander around complex cityscapes to deposit packages right outside customers' doors. Firms such as UPS and Wing have trialed urban rotary-wing drone delivery operations which were proven to be feasible and economic for rapid, on-demand delivery (Rejeb et al., 2021).

Drones with rotary-wing in Medicine Rotary-wing drones are employed in the transportation of emergency medicines including vaccines, blood samples, and medicine in remote or hard-to-reach areas. In the context of the COVID-19 outbreak, drones are instrumental in transporting medical products to quarantine areas, preventing human contact and enhancing response time (Euchi, 2020).

Rotary-wing drones are also employed in warehouse logistics for inventory management and internal transportation. They can autonomously scan barcodes, track inventory, and transport items within large warehouse facilities, improving operational efficiency and accuracy (Thomaidis & Zeimpekis, 2023).

4.3.3 Hybrid Drones

Hybrid drones represent a convergence of fixed-wing and rotary-wing technologies, offering a versatile solution that leverages the advantages of both types. These drones can transition between long-range, energy-efficient flights and precise, controlled vertical landings, making them suitable for a wide range of logistical scenarios. Their adaptability allows for efficient operations across various environments, from large-scale agricultural surveys to urban package deliveries. However, the integration of dual flight modes results in a more complex design, which can increase production costs and maintenance demands. The technical challenges associated with seamlessly combining these modes also require advanced control systems and expertise, potentially elevating the overall operational complexity.

4.3.3.1 Use Cases and Applications in Logistics

Hybrid drones are well suited for missions that require both long-range flight and precise vertical landing capabilities. They are used in logistics for delivering packages to areas where traditional fixed-wing drones cannot land, such as urban rooftops or constrained spaces. By combining the efficiency of fixed-wing flight with the versatility of rotary-wing drones, hybrid drones can cover greater distances while maintaining the ability to land in tight spots (Rejeb et al., 2021).

In the agriculture sector, hybrid drones are employed for precision farming tasks, such as spraying crops over large areas and then landing precisely to recharge or refill supplies. This capability enhances operational efficiency and reduces downtime (Godbole et al., 2019).

4.3.4 Advantages and Disadvantages of Each Type in Delivery Logistics

The table above provides a comparative overview of fixed-wing, rotary-wing, and hybrid drones, highlighting the distinct advantages and disadvantages of each type. Fixed-wing drones excel in long-range and high-speed missions but require specific infrastructure for takeoff and landing, limiting their flexibility. Rotary-wing drones, with their VTOL capabilities, offer unparalleled maneuverability and are well suited for urban environments, though they are constrained by shorter flight endurance and payload limitations. Hybrid drones emerge as a versatile option, integrating the strengths of both types, but their complex design and higher operational demands can pose challenges in terms of cost and maintenance. Each drone type presents unique trade-offs, making the choice dependent on the specific requirements of the application at hand (Hassanalian & Abdelkefi, 2017). The choice of drone type in

Type	Advantages	Disadvantages
Fixed-Wing Drones	Long-range, high-speed flight	Requires runways or catapult systems
	Energy-efficient, extended flight times	Limited maneuverability in confined spaces
	Ideal for large-area coverage and long-distance deliveries	Higher operational complexity and maintenance needs
Rotary-Wing Drones	Vertical takeoff and landing	Shorter flight range and speed
	High maneuverability and stability	Higher energy consumption, shorter battery life
	Suitable for urban logistics and last-mile deliveries	Limited payload capacity
Hybrid Drones	Combines fixed-wing and rotary-wing benefits	Complex design, higher cost
	Versatile for long-range and precise vertical operations	Higher maintenance requirements
	Efficient in diverse scenarios	Technical challenges in mode integration

FIGURE 4.1 Advantages and disadvantages of each type in delivery logistics

delivery logistics depends on specific mission requirements, operational environment, and logistical challenges. Because each type has unique benefits and drawbacks, they can be used in a variety of logistics and delivery-related applications.

4.4 TECHNOLOGICAL ADVANCEMENTS IN DELIVERY DRONES

Technological innovation in delivery drones has significantly influenced their efficiency, safety, and overall reliability in the logistics sector. This progress includes enhancements in navigation systems, control mechanisms, and power solutions, all contributing to the capacity of drones to operate in various environments with high precision. The integration of advanced technology allows drones to deliver packages efficiently, even in densely populated or challenging terrains.

4.4.1 Navigation and Control Systems for Logistics

Effective navigation and control systems are crucial for the operation of delivery drones. These systems enable drones to execute complex flight paths, maintain stability, and ensure safe passage to their delivery points. The development of these systems relies on the use of advanced technology, including GPS and obstacle detection mechanisms.

4.4.1.1 GPS and Satellite Navigation in Delivery

The use of Global Positioning System (GPS) and satellite navigation is fundamental for delivery drones. These location systems provide precise GPS positioning, allowing drones to follow complex routes, avoid ground obstacles, and precisely reach their home positions. With GPS technology, drones maintain their course during deliveries, quickly correct errors, and fly predictably. One of the major milestones in GPS technology has been utilizing Real-Time Kinematic (RTK) systems, improving the precision of standard GPS from several meters to sub-centimeter levels. This degree of accuracy is essential for delivery drones navigating densely populated urban areas, where exact landing and navigation are critical (Li et al., 2023). Additionally, multi-frequency GPS receivers help reduce signal interferences and errors caused by atmospheric conditions, enhancing reliability.

4.4.1.2 Obstacle Detection and Avoidance in Logistics

An essential ability of delivery drones is avoiding obstacles, which allow them to traverse the environment in a safe manner. State-of-the-art drones integrate ultrasonic, infrared, LiDAR, and visual cameras sensors to perceive and react to the environment.

Detailed 3D maps of the drone's surroundings are created using LiDAR technology, which measures distances with laser pulses. This helps drones to sense and avoid obstacles like trees, buildings, and power lines. Visual sensors, in conjunction with computer vision algorithms, enable a drone to identify and avoid dynamic obstacles, such as pedestrians and cars (Saunders et al., 2021).

These sensor technologies are backed up by complex algorithms for sensor data processing and real-time decision making. For example, the movement of obstacles can be predicted by means of machine learning algorithms and the drone can adapt its path accordingly (Macrina et al., 2020), thus improving safety and efficiency in multiple delivery schemes.

4.4.2 Battery and Power Systems for Delivery Drones

The success and efficiency of delivery drones heavily rely on their power systems, which impact their flight time, range, and payload capacity. As drone technology evolves, advancements in battery and power management systems are crucial to support the growing demands of logistics and delivery services. This section explores the types of batteries commonly used in delivery drones and highlights the recent innovations that enhance their performance.

4.4.2.1 Types of Batteries Used in Logistics Drones

The battery is what determines how a drone will perform with regard to flight time, range, etc. Lithium-polymer (Li-Po) or lithium-ion (Li-ion) batteries, well recognized for their high energy density and low mass, are the most common ones used in delivery drones.

Li-Po batteries are preferred because they can provide a high discharge rate that is necessary for the high power needs of takeoff and maneuvers. However, Li-ion batteries have a higher energy density and longer cycle life and are more appropriate for long-term missions. Battery chemistry and battery management systems are continually improved to improve performances and safety of these batteries (Euchi, 2020).

4.4.2.2 Innovations in Battery Technology for Delivery

The goals of recent advancements in battery technology are to improve charging times, decrease weight, and increase energy density. The adoption of solid-state batteries, which substitute a solid electrolyte for a liquid one and increase energy density and safety by lowering the possibility of leaks and fires, is one exciting advancement.

Another area of advancement is wireless charging and battery swapping systems. Wireless charging pads can be installed at delivery hubs, allowing drones to recharge without manual intervention. Battery swapping systems enable drones to quickly exchange depleted batteries for fully charged ones, significantly reducing downtime and increasing operational efficiency (Oigbochie et al., 2021).

4.4.3 Payload Capacity and Handling for Logistics

The payload capacity and handling mechanisms of delivery drones are vital to their performance in logistics operations. These factors directly influence the range, efficiency, and safety of drone deliveries. To maximize operational effectiveness, it is crucial to balance the weight distribution, secure the cargo appropriately, and leverage advanced engineering techniques.

4.4.3.1 Weight Considerations for Delivery Drones

The weight a delivery drone can carry is one of the most important factors that affects how far and efficiently it can fly. Flight time depends on many things: the weight of the drone, battery, sensors, and how it's built—all of which must match the payload to ensure good performance.

In order to combat weight, the body of the drone is constructed from lightweight materials like carbon fiber and high strength polymers. Recent advancements in materials science have introducéd composites that possess high strength-to-weight ratios, which would enable drone to carry more loads without loss of performance during flight (Korkmaz & Mohammed, 2020).

4.4.3.2 Package Securing Mechanisms for Logistics

Ensuring the safe and secure transport of packages is crucial for delivery drones. Various securing mechanisms have been developed to hold packages firmly during flight and release them accurately at the delivery point. These mechanisms include robotic grippers, magnetic clamps, and custom-designed cargo bays.

Robotic grippers, inspired by biological systems, can adapt to different package shapes and sizes, providing a versatile solution for various delivery scenarios. Magnetic clamps offer a quick-release mechanism, enabling rapid deployment of packages without manual intervention. Custom-designed cargo bays with cushioning and stabilization features ensure that delicate items are protected from vibrations and shocks during transit (Braun et al., 2019)

4.4.4 Autonomous vs. Piloted Drones in Delivery

The debate between autonomous and piloted drones continues to shape the landscape of drone technology in logistics. Both approaches offer unique advantages and are suitable for different types of delivery operations. The choice between autonomous and piloted systems depends on the specific demands of the delivery scenario, balancing automation and human expertise.

4.4.4.1 Autonomous Drone Technology in Logistics

Autonomous drones are equipped with advanced navigation systems, sensors, and artificial intelligence (AI) algorithms that enable them to operate without human intervention. These drones can plan their routes, avoid obstacles, and make real-time decisions based on environmental data. The autonomy level ranges from basic waypoint navigation to fully autonomous operations where the drone can adapt to dynamic changes in the environment.

One of the key benefits of autonomous drones is their ability to perform repetitive tasks with high precision and efficiency. In logistics, this translates to consistent and reliable delivery services, reducing human error and operational costs. Autonomous drones are particularly useful for last-mile deliveries, where they can navigate complex urban environments to reach their destinations (Li et al., 2023).

4.4.4.2 Advantages of Piloted Drones in Delivery

While autonomous drones offer significant advantages, piloted drones still play a crucial role in delivery logistics, especially in scenarios requiring human judgment and flexibility. Piloted drones are controlled by remote operators who can make real-time decisions based on visual and sensor data. This capability is essential in complex or unpredictable environments where autonomous systems may struggle to adapt.

Piloted drones are particularly valuable in emergency response and medical deliveries, where quick decision-making and adaptability are critical. For instance, in disaster-stricken areas, human operators can navigate through debris and challenging conditions to deliver medical supplies or conduct search and rescue operations (Bhatt et al., 2018).

The technological advancements in navigation and control systems, battery and power systems, payload capacity, and the balance between autonomous and piloted drones are driving the evolution of delivery drones in logistics. These innovations enhance the efficiency, safety, and reliability of drone-based delivery systems, making them a viable solution for modern logistics challenges.

4.5 APPLICATIONS IN DELIVERY AND LOGISTICS

Drones have revolutionized the delivery and logistics industry, offering solutions that address long-standing and emerging challenges. Their diverse applications span urban and rural deliveries, medical supply transportation, e-commerce logistics, and industrial operations. By integrating drone technology, logistics services can enhance speed, efficiency, and overall service quality while reducing operational costs and environmental impacts. The following sections explore these applications in greater detail, demonstrating the transformative potential of drones in the logistics sector.

4.5.1 Urban and Rural Deliveries

The use of drones in both urban and rural deliveries has demonstrated significant potential to enhance logistics by overcoming unique challenges associated with each environment. In urban settings, drones address issues such as congestion and limited vehicle access, while in rural areas, they bridge gaps in infrastructure and long-distance transport.

4.5.1.1 Benefits for Urban Logistics

Urban logistics present unique challenges due to dense populations, heavy traffic, and limited space for delivery vehicles. Drones offer significant advantages in this environment by providing rapid, flexible, and cost-effective delivery solutions. They can navigate through congested areas, bypass traffic jams, and deliver packages directly to customer locations, often within minutes.

Drones are particularly effective for last-mile deliveries, where traditional delivery vehicles might struggle with parking and access. Companies like Amazon and UPS have implemented drone delivery systems to improve delivery speed and efficiency in urban areas. For example, Amazon's Prime Air aims to deliver packages within 30 minutes using autonomous drones (Benarbia & Kyamakya, 2021).

Moreover, drones reduce the environmental impact of delivery services. They consume less energy compared to conventional delivery vehicles and contribute to lower carbon emissions, aligning with sustainability goals. By integrating drones into urban logistics, companies can enhance operational efficiency while minimizing their ecological footprint (Rejeb et al., 2021).

4.5.1.2 Challenges in Rural Deliveries

While drones offer significant benefits for urban logistics, they also hold great potential for rural deliveries. Rural areas often face challenges such as poor infrastructure, long distances, and limited access to essential services. Drones can bridge these gaps by providing direct, fast, and reliable delivery services to remote locations.

One notable example is Zipline, a company that uses drones to deliver medical supplies to rural and hard-to-reach areas in Rwanda and Ghana. These drones can cover long distances quickly, delivering critical medical supplies like blood and vaccines, thereby saving lives and improving healthcare outcomes (Law et al., 2023).

However, rural deliveries also present unique challenges, such as limited connectivity for GPS and communication systems, weather-related issues, and regulatory hurdles. Ensuring reliable connectivity and navigation in remote areas is crucial for the successful deployment of drone delivery systems. Addressing these challenges requires innovative solutions and collaborations with local authorities and communities to establish effective and sustainable delivery networks (Euchi, 2020).

4.5.2 Medical Supply and Emergency Response Deliveries

Drones have become crucial tools in medical logistics and emergency response due to their ability to deliver supplies rapidly and navigate challenging environments. Their use has enhanced healthcare delivery and emergency operations by providing timely, efficient solutions where traditional transportation methods fall short.

4.5.2.1 Delivery of Medical Supplies via Drones

Transport of medical supplies by drones has received much attention, especially in emergencies and pandemics. Drones have the potential to rapidly and efficiently deliver essential medical commodities, such as vaccines, drugs, and diagnostic specimens to health facilities and other areas, including areas that are difficult to reach. This ability is particularly important in situations where standard means of transportation are infeasible or too slow.

Throughout the COVID-19 pandemic, drones were crucial in transporting medical supplies to containment zones and hospitals, eliminating human interaction and expedite the delivery process. Aerial drones have been used to transfer blood samples and other urgent medical materials to serve immediate transport needs as well as minimize chances of contamination (Oigbochie et al., 2021).

4.5.2.2 Use in Disaster Response Logistics

In disaster response scenarios, drones provide a vital means of delivering supplies and conducting assessments. They can access areas that are otherwise unreachable due to damaged infrastructure, providing a lifeline for affected populations. Drones equipped with cameras and sensors can also conduct aerial surveys, helping disaster response teams to assess damage and plan relief efforts more effectively.

For example, during natural disasters such as hurricanes or earthquakes, drones can deliver food, water, and medical supplies to stranded individuals. They can also provide real-time data on the extent of damage and the locations of survivors, significantly improving the efficiency of rescue operations (Braun et al., 2019).

4.5.3 E-commerce and Retail Deliveries

The integration of drone technology into e-commerce and retail logistics has reshaped how goods are delivered, setting new benchmarks for speed, cost-efficiency, and customer satisfaction. Major e-commerce companies have adopted drones to optimize last-mile delivery, offering consumers faster and more reliable service.

4.5.3.1 Major E-commerce Implementations of Drones

E-commerce giants like Amazon and Alibaba have been at the forefront of integrating drones into their delivery networks. These companies aim to reduce delivery times and costs, enhance customer satisfaction, and gain a competitive edge by leveraging drone technology.

Amazon's Prime Air, for instance, has been testing drone deliveries to ensure packages reach customers within 30 minutes of ordering. Similarly, Alibaba's Cainiao Network has experimented with drone deliveries to remote areas in China, showcasing the potential of drones to revolutionize the e-commerce delivery landscape (Benarbia & Kyamakya, 2021).

4.5.3.2 Customer Experiences and Feedback in Logistics

Customer feedback on drone deliveries has been generally positive, highlighting the speed, convenience, and novelty of receiving packages via drones. Surveys and pilot programs have shown that customers appreciate the quick delivery times and the ability to track their packages in real time. However, there are concerns about privacy, safety, and the reliability of drone deliveries in adverse weather conditions (Li et al., 2023).

To address these concerns, companies are investing in robust safety protocols, reliable navigation systems, and customer education programs to ensure smooth and secure drone delivery experiences. Building trust and addressing public concerns are essential for the widespread adoption of drone deliveries in the e-commerce sector (Rejeb et al., 2021).

4.5.4 Industrial and Warehouse Logistics

Drones are transforming industrial and warehouse logistics by automating tasks, enhancing productivity, and optimizing operations. Their integration into warehouses and industrial facilities provides a new level of efficiency and accuracy, streamlining processes that traditionally relied on manual labor and time-intensive practices.

4.5.4.1 Inventory Management Using Drones

In industrial and warehouse logistics, drones are being used to automate inventory management processes. Equipped with RFID scanners and cameras, drones can quickly scan barcodes and inventory tags, providing real-time data on stock levels, locations, and conditions. This automation reduces the need for manual inventory checks, increasing efficiency and accuracy.

Drones can navigate through warehouse aisles, reaching high shelves and hard-to-access areas, performing inventory checks faster than human workers. This capability not only speeds up the inventory management process but also reduces the risk of human error and improves overall operational efficiency (Korkmaz & Mohammed, 2020).

4.5.4.2 Internal Logistics Within Warehouses

Drone applications also include internal warehouse logistics, for example, moving small products, papers, and specimens within multiple sections of the building. By using drones in this task, operations are streamlined, human labor reduced, and productivity improved.

Automating warehouse operations through the incorporation of drones in warehouse management systems ensures that internal logistics processes are optimized, inventory accuracy is increased, and costs are lowered. Drones' ability to adapt and offer flexibility positions them as the perfect fit for dynamic and complex warehouse settings (Nyaaba & Ayamga, 2021).

Drone uses in logistics and delivery are diverse and revolutionary. Drones offer innovative solutions to boost productivity and efficiency, reduce costs, and enhance the quality of service in many industries, which range from medical supply to e-commerce as well as warehouse management and urban and rural delivery. Drones are expected to play a vital role in the logistics sector as the world advances in terms of technological developments.

4.6 MATHEMATICAL MODELS AND ALGORITHMS IN DRONE DELIVERY AND LOGISTICS

The successful integration of drones into logistics relies heavily on mathematical models and algorithms that enable efficient operations. These models address key aspects such as route optimization, energy management, collision avoidance, and material selection. By applying these mathematical principles, drone delivery systems can achieve optimal performance, ensuring timely, safe, and cost-effective solutions. This chapter outlines core mathematical concepts and algorithms fundamental to drone logistics, linking these theoretical models with their practical applications (Benarbia & Kyamakya, 2021).

4.6.1 Mathematical Models in Drone Logistics

The integration of drones into logistics requires a deep understanding of the mathematical models and algorithms that drive efficient operations. These models are essential for solving complex problems related to route optimization, energy management, collision avoidance, and material selection. By leveraging mathematics, drone delivery systems can achieve optimal performance, ensuring timely and cost-effective logistics solutions. This chapter delves into the core mathematical concepts and algorithms that underpin drone logistics, connecting these models with the practical applications discussed in previous sections (Benarbia & Kyamakya, 2021).

4.6.2 Optimization Algorithms for Route Planning

Efficient route planning is crucial for minimizing delivery time, energy consumption, and operational costs. Mathematical optimization models such as the Traveling Salesman Problem (TSP), the Vehicle Routing Problem (VRP), and other advanced algorithms like A* Search and Dijkstra's Algorithm are widely used in drone logistics to solve these challenges (Rejeb et al., 2021).

4.6.2.1 Traveling Salesman Problem (TSP)

The TSP is a classic optimization problem that seeks to find the shortest possible route for a drone to visit a given set of locations exactly once and return to the starting point. This problem is vital in logistics, where minimizing travel distance directly translates to reduced energy consumption and faster deliveries (Rejeb et al., 2021).

The solution to the TSP must satisfy constraints ensuring that each location is visited exactly once and that the route forms a single tour. Given the NP-hard nature of the TSP, heuristic methods like Genetic Algorithms and Simulated Annealing are often employed for practical solutions.

4.6.2.2 Vehicle Routing Problem (VRP)

The VRP generalizes the TSP by taking into consideration a fleet of drones that have a capacity limit. The problem involves finding the best set of routes that have the

Mathematically, the TSP can be expressed as:

$$\text{Minimize} \sum_{i=1}^{n}\sum_{j=1}^{n} d(i,j)\cdot x_{ij}$$

Where:

- $d(i,j)$ is the distance between locations i and j,
- x_{ij} is a binary variable indicating whether the route includes a direct path from i to j.

FIGURE 4.2 Equation of traveling salesman problem (TSP)

lowest total cost of delivery without missing a single delivery within the capacity and

The VRP can be formulated as:

$$\text{Minimize} \sum_{k=1}^{m}\sum_{i=1}^{n}\sum_{j=1}^{n} d(i,j)\cdot x_{ijk}$$

Where:

- x_{ijk} indicates whether drone k travels directly from location i to j,
- m is the number of drones,
- Q_k is the capacity of drone k.

FIGURE 4.3 Equation of vehicle routing problem (VRP)

time limit of the drones (Rejeb et al., 2021).

Scheduling the multiple drones concurrently to solve the VRP can be achieved using techniques such as Mixed-Integer Linear Programming or metaheuristic ones that include Tabu Search and Particle Swarm Optimization (Rejeb et al., 2021).

4.6.2.3 A* Search Algorithm

The A* Search Algorithm is widely used in pathfinding and graph traversal. It helps in finding the most efficient path between two points, which is particularly useful for drone navigation in dynamic environments. A* uses a combination of heuristics to estimate the cost to reach the goal from a given node, making it faster than other exhaustive search algorithms (Rejeb et al., 2021).

The algorithm works as follows:

1. Start at the initial node and add it to an open list.
2. Calculate the cost function f(n) = g(n) + h(n) for each node, where:
   - g(n) is the cost to reach the node from the start.
   - h(n) is the heuristic estimate of the cost to reach the goal from the node.

3. Expand the node with the lowest f(n) value, and move it to the closed list.
4. Repeat until the goal is reached.

This algorithm is particularly useful for real-time route adjustments, especially in environments where obstacles and other drones are dynamically changing (Rejeb et al., 2021).

4.6.2.4 Dijkstra's Algorithm

Dijkstra's Algorithm is another fundamental algorithm used for finding the shortest paths between nodes in a graph, which represents a map for drone delivery. Unlike A*, Dijkstra's algorithm doesn't use a heuristic and guarantees the shortest path by expanding the least costly node first (Rejeb et al., 2021).

The algorithm proceeds with the following steps:

1. Initialize the starting node with a tentative distance of 0 and all other nodes with infinity.
2. Set the current node as the starting node and mark all others as unvisited.
3. For the current node, calculate the tentative distance to each neighboring node.
4. If the calculated distance of a node is less than the known distance, update the distance.
5. Mark the current node as visited, and select the unvisited node with the smallest tentative distance as the new current node.
6. Repeat until all nodes are visited or the shortest path to the destination is found.

This algorithm is especially effective in dense environments where there are many potential paths, ensuring drones always take the most efficient route (Rejeb et al., 2021).

4.6.3 Energy Consumption Models in Drone Logistics

Energy efficiency is a critical factor in drone operations, directly affecting the range, payload capacity, and operational costs. Mathematical models are used to estimate and optimize energy consumption during drone flights (Rejeb et al., 2021).

4.6.3.1 Basic Energy Consumption Model

By minimizing energy consumption through optimized flight paths and speeds, drones can extend their operational range and reduce costs (Rejeb et al., 2021).

4.6.3.2 Energy Consumption with Payload

This model helps in planning deliveries by accounting for the additional energy required to transport goods, ensuring that drones do not exceed their energy capacity (Rejeb et al., 2021).

The basic energy consumption model for a drone can be represented as:

$$E = P \times t$$

Where:

- E is the total energy consumed,
- P is the power required for flight, dependent on factors such as speed and drag,
- t is the flight time.

The power P can be further detailed as:

$$P = \frac{1}{2} C_d \rho A v^3$$

Where:

- C_d is the drag coefficient,
- ρ is the air density,
- A is the cross-sectional area,
- v is the velocity.

FIGURE 4.4 Equations of energy consumption and power

Carrying a payload increases energy consumption, which can be modeled as:

$$E_{\text{payload}} = E_{\text{no payload}} + m_{\text{payload}} \cdot g \cdot d \cdot \text{efficiency factor}$$

Where:

- m_{payload} is the mass of the payload,
- g is the acceleration due to gravity,
- d is the distance traveled.

FIGURE 4.5 Equation of energy consumption with payload

4.6.4 Collision Avoidance Algorithms

Collision avoidance is vital for the safe operation of drones, particularly in complex environments with dynamic obstacles. Mathematical algorithms are used to enable real-time navigation and obstacle avoidance (Saunders et al., 2021).

4.6.4.1 Potential Field Method

The Potential Field Method is a popular approach for real-time collision avoidance. In this method, drones are influenced by virtual forces that guide them toward the target while avoiding obstacles.

This method ensures that the drone navigates safely to its destination, avoiding collisions with obstacles (Saunders et al., 2021).

The overall force $\vec{F}$ acting on the drone is the sum of attractive forces towards the goal and repulsive forces from obstacles:

$$\vec{F} = \vec{F}_{\text{attractive}} + \sum_{i=1}^{k} \vec{F}_{\text{repulsive},i}$$

Where:

- $\vec{F}_{\text{attractive}} = -k_a \cdot (\vec{x} - \vec{x}_{\text{goal}})$ is the attractive force.
- $\vec{F}_{\text{repulsive},i} = \frac{k_r}{\|\vec{x} - \vec{x}_{\text{obstacle},i}\|^2} \cdot \frac{\vec{x} - \vec{x}_{\text{obstacle},i}}{\|\vec{x} - \vec{x}_{\text{obstacle},i}\|}$ is the repulsive force.

FIGURE 4.6 Equation of potential field method

4.6.4.2 Rapidly Exploring Random Tree (RRT)

Rapidly exploring Random Tree (RRT) is an algorithm used for path planning that efficiently searches high-dimensional spaces by randomly building a space-filling tree. This is particularly useful for drones operating in environments with unpredictable or moving obstacles (Saunders et al., 2021).

The algorithm works as follows:

1. Start at an initial position and build a tree by randomly selecting a point in the space.
2. Extend the tree toward the randomly selected point by connecting it to the nearest node in the tree.
3. If the tree can connect to the goal without encountering obstacles, a path is found.
4. If not, continue growing the tree by selecting new random points until a valid path is found or the tree fully explores the space.

RRT is effective for real-time pathfinding because it does not require a pre-computed map and can dynamically adapt to new obstacles (Saunders et al., 2021).

4.6.4.3 Kalman Filter for Navigation

A recursive method called the Kalman Filter uses a sequence of noisy and insufficient observations to estimate the state of a dynamic system. It is widely used in drone navigation to filter errors from sensor data and provide accurate real-time position estimates (Saunders et al., 2021).

The Kalman Filter operates in two phases:

1. Prediction Phase: Predict the current state and uncertainty using the system's previous state and known control inputs.
2. Update Phase: Adjust the predicted state using the actual measurement and update the uncertainty to reflect the new state estimate.

This filtering process allows drones to maintain stable flight and accurate navigation even in the presence of disturbances and sensor noise (Saunders et al., 2021).

4.6.5 Material Selection Models for Drone Design

Material selection is crucial for optimizing the performance of drones, balancing weight, strength, and cost. Mathematical models are used to determine the best materials for various drone components (Rejeb et al., 2021).

4.6.5.1 Multi-objective Optimization for Material Selection

The selection of materials can be approached as a multi-objective optimization problem, where the goal is to minimize weight and cost while maximizing strength.

This model helps in selecting materials that optimize the drone's performance, ensuring it can carry the required payloads while maintaining structural integrity and minimizing costs (Rejeb et al., 2021).

4.6.6 Application of Mathematical Models in Real-World Scenarios

The algorithms and mathematical models presented in the chapter are utilized in a range of real-life logistics applications for drones, including the optimization of delivery paths as well as the improvement of energy efficiency and safety in navigation. Logistics firms benefit from incorporating these models into their operations by realizing substantial advances in efficiency, cost-cutting, and quality of service (Benarbia & Kyamakya, 2021).

For instance, Amazon and UPS use optimization algorithms to optimize their drone delivery routes to reduce delivery times and consumption of energy(Benarbia & Kyamakya, 2021). In the same way, consumption models for energy are employed to schedule flights to optimize the flying range of drones to successfully complete their deliveries without draining their power (Benarbia & Kyamakya, 2021).

Material selection models in warehouse logistics aid in the designing of light but robust drones that can support repetitive flights within huge structures. In the urban environment, collision avoidance algorithms play an important role in which drones have to navigate safely through congested airspaces(Saunders et al., 2021).

$$\text{Minimize } f(\vec{x}) = w_1 \cdot \text{Weight}(\vec{x}) + w_2 \cdot \frac{1}{\text{Strength}(\vec{x})} + w_3 \cdot \text{Cost}(\vec{x})$$

Subject to:

- Maximum allowable stress:

$$\sigma_{\text{material}} \leq \sigma_{\text{allowable}}$$

- Maximum allowable deflection:

$$\delta_{\text{material}} \leq \delta_{\text{allowable}}$$

Where:

- $\vec{x}$ represents the design variables (e.g., thickness, material type).
- w_1, w_2, w_3 are weighting factors representing the importance of each objective.

FIGURE 4.7 Equation of multi-objective optimization for material selection

In aggregate, these mathematical models form the basis for the effective integration of drones in contemporary logistics to facilitate the creation of creative solutions to keep up with the needs of today's high-speed delivery networks (Rejeb et al., 2021).

Mathematics has a central role in the use of drones for delivery and logistics as it offers the mathematical tools and frameworks to maximize the efficiency of the operations, effectively allocate resources, and guarantee safety. The mathematical models and algorithms in this chapter illustrate how mathematical concepts are used for the resolution of intricate logistical problems in order to pave the way for future advanced and extensive application of drones. With advancements in technology, these mathematical models will also continue to improve, promoting further developments in the application of drone logistics (Rejeb et al., 2021).

4.7 BENEFITS OF DRONES IN LOGISTICS

The integration of drones into logistics has brought significant advancements across various aspects of delivery operations. Drones offer unique benefits that include improving speed and efficiency, reducing operational costs, enhancing environmental sustainability, and increasing accessibility to remote areas.

4.7.1 Speed and Efficiency Improvements

Drones improve the speed and logistical efficiency of operations such as last-mile delivery. Traffic congestion, road conditions, and multiple deliveries can often create significant obstacles for traditional distribution methods. Drones, however, can bypass these obstructions, traveling directly to the delivery point and significantly reducing routing times. This advantage is particularly useful in urban areas where traffic congestion can cause extensive delays.

Studies indicate that drones can cut delivery times by half, making them a compelling choice for companies aiming to offer faster delivery services. For example, Amazon Prime Air's drone delivery solution promises to drop off packages within 30 minutes of purchase, highlighting the potential for rapid service (Benarbia & Kyamakya, 2021).

Drones can work non-stop around the clock without requiring breaks like human drivers, greatly enhancing operational efficiency. This ability is especially valuable during peak times, like the holiday season, when quick deliveries become crucial for meeting heightened demand (Rejeb et al., 2021).

4.7.2 Cost Reductions in Delivery Services

Drones offer substantial cost-saving opportunities in logistics by reducing labor and operational costs. The automation of delivery processes minimizes the need for human labor, which is one of the highest cost components in logistics. Moreover, drones can cover multiple delivery points in a single flight, reducing fuel and maintenance costs associated with traditional delivery vehicles.

A comprehensive cost-benefit analysis revealed that drones could lower delivery costs by up to 40% compared to traditional delivery methods. These savings are achieved through reduced fuel consumption, lower maintenance costs, and decreased labor expenses (Raj & Sah, 2019). The use of electric drones contributes to lower energy costs, making them a more sustainable and economical choice for logistics operations.

4.7.3 Environmental Impact and Sustainability in Logistics

Drones in logistics can bring substantial environmental benefits. Electric-battery-powered drones produce zero emissions during operation, making them a greener alternative to traditional delivery vehicles that rely on fossil fuels. This shift is especially relevant with increasing regulatory and corporate sustainability mandates focused on reducing carbon emissions.

Studies show that drones can cut greenhouse gas emissions from logistics operations by 30% or more, achieved through lower energy consumption and the reduction of idle time associated with traffic congestion (Euchi, 2020).

Drones can operate on renewable energy, such as solar-powered charging stations, which enhances their sustainability credentials further. This capability aligns with the global push for clean energy transitions and greenhouse gas emission reductions (Rejeb et al., 2021).

4.7.4 Enhanced Accessibility to Remote Areas Through Drones

Drones excel in providing delivery services to remote and hard-to-reach areas. Traditional delivery vehicles often face challenges such as poor road infrastructure, long travel times, and accessibility issues. Drones can overcome these barriers by flying directly to the destination, making them ideal for delivering essential goods to remote locations.

For example, in healthcare logistics, drones have been used to deliver medical supplies, vaccines, and blood samples to rural and underserved areas, significantly improving access to critical healthcare services. Zipline, a drone delivery service, has successfully implemented this model in Rwanda and Ghana, reducing delivery times from hours to minutes and ensuring timely access to medical supplies (Law et al., 2023).

Moreover, drones are instrumental in disaster response scenarios, where they can quickly deliver supplies to affected areas, conduct aerial surveys, and support rescue operations. Their ability to operate in challenging environments enhances their utility in emergency logistics (Braun et al., 2019).

The benefits of drones in logistics are manifold, ranging from speed and efficiency improvements to cost reductions, environmental sustainability, and enhanced accessibility to remote areas. These advantages make drones a transformative technology in the logistics sector, driving innovation and efficiency in delivery services.

4.8 CHALLENGES AND LIMITATIONS IN DELIVERY LOGISTICS

While the integration of drones into delivery logistics holds significant promise, several challenges and limitations must be addressed for their widespread and effective use. These challenges include regulatory and legal barriers, safety and privacy concerns, technological constraints, and public perception issues. This chapter explores these critical areas and the measures needed to overcome them.

4.8.1 Regulatory and Legal Hurdles in Drone Delivery

Drone delivery regulations are complicated and differ significantly between jurisdictions. Drone operations may be severely hampered by current rules, especially when it comes to autonomous and long-distance delivery.

4.8.1.1 Current Regulations Affecting Logistics

The utilization of drones in delivery logistics is governed by stringent regulations, which considerably differ from one region to another. The U.S. Federal Aviation Administration and European Union Aviation Safety Agency (EASA) have written regulations to allow the incorporation of drones into domestic airspaces safely, as well. Typical regulations encompass restrictions on flight altitudes, no-fly zones, operator certifications, and a mandate for drones to always remain within the visual line-of-sight (VLOS) of operators (Rejeb et al., 2021).

Existing regulations can drastically restrict what drones are allowed to do, especially when it comes to distant or autonomous deliveries. The FAA's Part 107 rules, for example, enforce VLOS, which effectively means we won't be seeing drones used for deliveries over fairly short distances. In addition, there are significant restrictions on night operations and flights over people, which would further limit the use of drones in urban areas (Turgut & Şeker, 2023).

4.8.1.2 Potential Regulatory Changes for Delivery

As drone technology advances and its benefits become more apparent, there is ongoing advocacy for regulatory changes to accommodate more extensive drone operations. Proposed changes include the development of Unmanned Traffic Management (UTM) systems to facilitate safe and efficient airspace integration and the relaxation of VLOS requirements through the adoption of Beyond Visual Line of Sight (BVLOS) operations.

Regulatory bodies are also considering frameworks for certifying autonomous drones and establishing corridors for drone deliveries to ensure safe and efficient logistics operations. These changes aim to strike a balance between innovation and safety, enabling more widespread and effective use of drones in delivery logistics (Li et al., 2023).

4.8.2 Safety and Privacy Concerns in Logistics

The integration of drones into delivery logistics presents significant safety and privacy challenges, particularly as drones share airspace with other aircraft and operate over populated areas. Addressing these issues is essential to ensure safe operations and public trust.

4.8.2.1 Safety Issues in Drone Delivery

Safety is a paramount concern in the deployment of drones for delivery logistics. Potential hazards include mid-air collisions with other aircraft, crashes due to technical failures, and risks to people and property on the ground. To mitigate these risks, drones are equipped with advanced safety features such as collision avoidance systems, redundant flight controls, and emergency landing protocols.

Despite these advancements, the integration of drones into shared airspace remains a significant challenge. Ensuring reliable communication and coordination between drones and other airspace users is critical to prevent accidents. Research is ongoing to develop robust air traffic management systems and enhance the reliability of drone technologies (Rejeb et al., 2021).

4.8.2.2 Privacy Concerns and Public Perception in Logistics

There are serious privacy issues with the usage of drones for deliveries. Potential privacy violations may result from drones with cameras and sensors unintentionally taking pictures and data from private sites. Public perception of drones is often influenced by concerns over surveillance and data security.

To address these issues, companies must implement strict data protection policies and ensure transparency about how data collected by drones is used and stored. Engaging with communities to build trust and addressing their concerns proactively can also help improve public perception and acceptance of drone deliveries (Law et al., 2023).

4.8.3 Technological Limitations in Delivery

Technological challenges remain a significant hurdle for the widespread adoption and effectiveness of drone delivery. These limitations include constraints related to battery life, flight range, and environmental resilience.

4.8.3.1 Battery Life and Range Constraints

One of the big technological bottlenecks in delivery drones is battery life. Nearly all drones currently use lithium-polymer or lithium-ion batteries, with constraints on flight time and range that limit their viability for long-distance deliveries. This issue necessitates frequent recharging or battery swaps.

Battery technology advancements are under study, with solid-state batteries and hydrogen fuel cells showing potential to boost energy density and extend flight duration. Moreover, establishing effective recharging infrastructure—such as wireless charging pads and automated battery swap stations—can help reduce these limitations (Rejeb et al., 2021).

4.8.3.2 Weather and Environmental Factors Affecting Delivery

Drone operations face significant challenges with difficult weather. Strong winds, heavy rain, snow, or extreme temperatures can impact flight stability and safety. In areas with variable climates, adverse weather can limit the reliability and consistency of drone delivery services.

Research continues to develop drones with robust designs and advanced sensors capable of navigating challenging weather. Additionally, drones can benefit from real-time weather data integration and predictive analytics to plan the safest, most efficient routes (Raj & Sah, 2019).

4.8.4 Public Perception and Acceptance of Drone Logistics

Public perception and community acceptance are critical for the successful integration of drones in delivery logistics. While the technology offers clear benefits, public concerns related to safety, privacy, and noise can impact its broader acceptance and usage.

4.8.4.1 Community Reactions to Drone Delivery

For a service that relies on drones flying overhead, gaining public acceptance for drone deliveries is essential. Community responses can vary: some welcome the convenience and innovation that drones introduce, while others express concerns about noise, privacy, and safety. Building public trust requires effective communication, community engagement, and transparent operations to assure the public that their data and safety are prioritized.

Efforts like public consultations, demonstrations, and collaboration with local authorities help align drone operations with community standards. Highlighting tangible benefits, such as faster delivery times and reduced environmental impact, can also foster public support for drone use (Moshref-Javadi & Winkenbach, 2021).

4.8.4.2 Media Portrayal of Drone Logistics

Media portrayal plays a significant role in shaping public perception of drone logistics. Positive media coverage highlighting successful implementations, technological advancements, and the benefits of drone delivery can enhance public acceptance. Conversely, negative coverage focusing on accidents, privacy breaches, or regulatory violations can fuel public skepticism and resistance.

Companies and industry stakeholders need to engage with the media proactively, providing accurate information and addressing any misconceptions. Transparency about safety measures, regulatory compliance, and the societal benefits of drone technology can help create a more informed and supportive public opinion (Rejeb et al., 2021).

While drones offer transformative potential for logistics, several challenges and limitations must be addressed to realize their full benefits. Regulatory and legal hurdles, safety and privacy concerns, technological limitations, and public perception issues are critical areas that require ongoing attention and innovation. By addressing these challenges, the logistics industry can leverage drone technology to enhance efficiency, reduce costs, and improve service delivery.

4.9 CASE STUDIES IN DRONE DELIVERY

Real-world case studies offer invaluable insights into how drones are being utilized successfully in logistics and delivery, showcasing their potential benefits as well as the challenges faced during implementation. This section highlights key examples of how companies and government entities are leveraging drone technology to enhance operations and improve delivery outcomes.

4.9.1 Successful Implementations by Companies

Companies at the forefront of the logistics industry have been pioneering the use of drones to streamline delivery processes, optimize efficiency, and solve last-mile delivery issues. Two notable examples include Amazon Prime Air and UPS Flight Forward.

4.9.1.1 Amazon Prime Air's Logistics Operations

Amazon Prime Air is one of the most prominent examples of commercial drone delivery. Launched with the goal of delivering packages to customers within 30 minutes, Amazon Prime Air has conducted extensive testing and pilot programs to integrate drones into its logistics network. The drones used by Amazon are designed for autonomous operation, outfitted with advanced sensors and navigation systems to guarantee effective and safe delivery.

The benefits observed from Amazon's implementation include reduced delivery times, enhanced customer satisfaction, and lower operational costs. The use of drones has also helped Amazon address last-mile delivery challenges, particularly in urban areas where traffic congestion can delay deliveries (Benarbia & Kyamakya, 2021). However, regulatory hurdles and public acceptance remain significant challenges that Amazon continues to address through ongoing research and development efforts (Rejeb et al., 2021).

4.9.1.2 UPS Flight Forward's Delivery Services

UPS Flight Forward has made significant strides in integrating drones into its delivery services. In 2019, UPS received approval from the Federal Aviation Administration (FAA) to operate a "drone airline," allowing it to conduct beyond-visual-line-of-sight (BVLOS) operations. This milestone enabled UPS to expand its drone delivery operations, particularly in healthcare logistics.

UPS Flight Forward has partnered with healthcare facilities to deliver medical supplies, such as blood samples and medications, directly to hospitals and laboratories. This service has proven particularly valuable in emergencies and during the COVID-19 pandemic, where quick and contactless deliveries were essential. The success of

UPS Flight Forward underscores the potential of drones to revolutionize healthcare logistics by providing rapid, reliable, and safe delivery solutions (Law et al., 2023).

4.9.2 Government and Military Use Cases in Logistics

Drones have not only been embraced by private companies but have also found valuable applications in government and military logistics. Their use in these sectors underscores the versatility and strategic advantages of drone technology in high-stakes and specialized scenarios.

4.9.2.1 Military Logistics with Drones

The military has been at the forefront of drone technology development, leveraging UAVs for various logistics and operational tasks. Drones are used for reconnaissance, surveillance, and delivering supplies to troops in remote or hostile environments. The ability to operate in difficult terrains and perform autonomous missions makes drones an invaluable asset in military logistics.

One notable example is the use of drones by the U.S. military to deliver critical supplies to frontline troops in conflict zones. These drones can carry medical supplies, ammunition, and food, ensuring that soldiers receive the necessary resources without exposing human operators to danger. The success of military drone logistics has inspired similar applications in civilian sectors, highlighting the versatility and reliability of UAV technology (Rejeb et al., 2021).

4.9.2.2 Government-Sponsored Delivery Projects

Governments around the world are exploring the use of drones to enhance public services and improve logistics efficiency. One such example is the government of Rwanda, which has teamed up with drone delivery business Zipline to provide medical supplies to far-flung locations. This initiative has significantly reduced delivery times for critical supplies, such as blood and vaccines, from hours to minutes, improving healthcare outcomes in underserved regions (Law et al., 2023).

Similarly, the Australian government has implemented drone delivery projects to improve logistics in rural and isolated communities. These projects focus on delivering essential goods, such as groceries and medications, to areas with limited access to traditional delivery services. The success of these government-sponsored projects demonstrates the potential of drones to enhance public services and address logistical challenges in diverse environments (Euchi, 2020).

4.9.3 Lessons Learned from Early Adopters in Delivery Logistics

Early participants in the use of drone logistics have learned useful lessons about the pros and the cons of deploying UAVs in supply chain operations. These include the need for regulatory compliance, the importance of having strong safety procedures in place, and the value of public education to establish public acceptance and confidence.

Successful implementations have shown that drones can significantly enhance delivery speed, reduce costs, and improve service reliability. However, challenges such as regulatory restrictions, technological limitations, and public concerns about privacy and safety must be addressed to achieve widespread adoption.

For instance, Amazon and UPS have proved the value of working in conjunction with the authorities in order to secure the necessary permits and carry on their operations safely. They have also made significant investments in new-age technologies for enhanced performance and security of drones in the form of collision avoidance systems and autonomous navigation systems (Rejeb et al., 2021).

Public engagement and education are crucial for building trust and acceptance of drone delivery services. Companies that have successfully implemented drone logistics have conducted community outreach programs, provided transparent information about their operations, and addressed public concerns proactively. These efforts have helped mitigate resistance and foster positive perceptions of drone technology (Benarbia & Kyamakya, 2021).

The case studies of successful drone delivery implementations highlight the transformative potential of UAVs in logistics. From commercial ventures like Amazon Prime Air to government-sponsored projects and military applications, drones offer innovative solutions to complex logistical challenges. By learning from early adopters, the logistics industry can leverage these insights to drive future advancements and achieve broader adoption of drone technology.

4.10 FUTURE TRENDS AND PREDICTIONS IN DRONE DELIVERY

Drone delivery is expected to undergo significant development in the future thanks to new technologies and creative strategies. These developments have the potential to change supply chain dynamics and logistics, offering new capabilities that enhance efficiency, flexibility, and sustainability. From breakthroughs in AI and design to potential economic impacts, the future of drone logistics holds considerable potential for transformation and growth.

4.10.1 Emerging Technologies and Innovations in Logistics

As the field of drone delivery evolves, breakthroughs in technology and material science are setting the stage for significant transformations. These innovations are expected to enhance the operational capabilities of drones, improve efficiency, and open up new possibilities for logistics applications.

4.10.1.1 Advances in Drone AI for Delivery

Artificial Intelligence (AI) is poised to play a pivotal role in the future of drone delivery logistics. AI technologies enhance autonomous navigation, obstacle detection, and avoidance capabilities, allowing drones to operate safely and efficiently in complex environments. Machine learning algorithms enable drones to learn from their surroundings and improve their performance over time. For instance, AI can optimize flight paths to minimize energy consumption and avoid no-fly zones, making drone operations more reliable and cost-effective (Rejeb et al., 2021).

Moreover, AI-driven data analytics can provide valuable insights into operational efficiencies and customer preferences, allowing logistics companies to refine their services and enhance customer satisfaction. Predictive maintenance, powered by AI, can also preemptively identify potential mechanical issues, reducing downtime and extending the lifespan of drones.

4.10.1.2 New Materials and Designs for Logistics Drones

The development of new materials and innovative designs is set to revolutionize drone capabilities. Lightweight yet durable materials such as carbon fiber composites and advanced polymers can enhance drone performance by reducing weight and increasing payload capacity. These materials also improve the durability and resilience of drones, allowing them to withstand harsh environmental conditions.

Innovative design concepts, including foldable drones and modular components, are being explored to enhance flexibility and ease of transportation. Foldable drones can be compactly stored and deployed quickly, making them ideal for rapid response scenarios and urban logistics. Modular designs enable easy replacement of damaged parts, reducing maintenance costs and downtime (Benarbia & Kyamakya, 2021).

4.10.2 Potential Impact on Global Logistics from Drone Delivery

The integration of drones into logistics systems is expected to have profound implications for global supply chains and economic structures. This transformation could enhance delivery capabilities, reshape traditional supply chain models, and introduce new business opportunities.

4.10.2.1 Changes in Supply Chain Dynamics due to Drones

Autonomous drones will transform logistics operations and reshape the supply chain landscape as we now know it. Drones allow for faster, point-to-point delivery, supporting more decentralized and dynamic supply chains. This decentralization can decrease reliance on large distribution centers and offers opportunities for local inventory management, enabling companies to respond quickly to market demands and significantly reduce lead times.

Additionally, drones can serve as an effective last-mile delivery solution for e-commerce, ensuring quicker and more reliable deliveries from retailers to consumers. This capability can enhance customer satisfaction and drive e-commerce growth, especially in regions with underdeveloped logistics infrastructure (Moshref-Javadi & Winkenbach, 2021).

4.10.2.2 Long-Term Economic Effects of Drone Logistics

The long-term economic impact of drone logistics is substantial. Businesses adopting drones can significantly reduce their operational expenses and enhance delivery efficiency, leading to considerable cost savings. These benefits can, in turn, be shared with consumers through reduced prices, boosting competitiveness and fostering broader economic growth. Additionally, the widespread integration of drone technology is likely to create numerous job opportunities across sectors such as drone manufacturing, maintenance, and data analytics.

Drones also pave the way for innovative business models, notably drone-as-a-service (DaaS). Under this approach, companies can lease drones for specific logistic tasks, lowering the entry barrier for small and medium-sized enterprises (SMEs). This model enables SMEs to benefit from cutting-edge delivery technologies without the need for significant initial investments (Raj & Sah, 2019).

4.10.3 Predictions for the Next Decade in Delivery Logistics

The coming decade is expected to witness exponential growth in the adoption and integration of drones in delivery logistics. Several key trends and predictions include:

- Large-scale Commercial Application: With the advancement in regulation and the maturity of the technology itself, commercial use of drones for delivery purposes should ultimately become the norm for last-mile delivery and in urban spaces.
- Increased Autonomous Functions: Advances in machine learning and AI will lead to the creation of highly autonomous drones that will navigate and make decisions independently to a larger extent.
- Hybrid Delivery Models: Hybrid delivery models that incorporate drones along with conventional delivery vehicles will more commonly be utilized. It capitalizes on the best of both modes to maximize efficiency and reach.
- Sustainability Emphasis: Drones will assume a focal role in sustainable logistics procedures through the minimization of emissions of carbon and advancing environmentally friendly transport solutions. Their environmental benefit will be complemented by the use of renewable energy, pertaining to solar-operating drones.
- Global Outreach: The use of drones in logistics will spread to developing economies to overcome infrastructural limitations and increase the availability of goods and services in rural communities (Rejeb et al., 2021).

Drone delivery logistics in the future is characterized by the speedy evolution of technology, advancements in regulation, and new business models. These trends will reshape supply chain dynamics, increase logistics efficiency worldwide, and support sustainable economic development.

4.11 CONCLUSION

4.11.1 Recap of Key Points in Drone Delivery Logistics

In the logistics industry, drones have become a game-changing technology with a host of advantages, including faster delivery, lower costs, improved accessibility, and environmental sustainability. Drone integration into logistics operations has shown great promise for a number of uses, such as e-commerce, warehouse management, medical supply transportation, and delivery in both urban and rural areas.

- Speed and Efficiency: By avoiding conventional transportation issues like traffic jams and inadequate infrastructure, drones increase delivery speed and operational efficiency. They offer quick, on-demand service and are especially efficient at last-mile delivery (Rejeb et al., 2021).
- Cost Reductions: By automating delivery processes and reducing reliance on human labor and fuel consumption, drones contribute to substantial cost savings in logistics operations (Raj & Sah, 2019).
- Environmental Impact: Drones offer a greener alternative to traditional delivery methods, significantly reducing carbon emissions and promoting sustainable logistics practices (Euchi, 2020).
- Accessibility: Drones are capable of reaching remote and hard-to-access areas, making them invaluable for delivering critical supplies in rural and disaster-stricken regions (Law et al., 2023).

4.11.2 The Potential of Drones to Revolutionize Delivery Services

The potential of drones to revolutionize delivery services is immense, driven by continuous advancements in technology, regulatory support, and innovative business models. Key areas of impact include:

- Technological Advancements: Innovations in AI, materials science, and battery technology will enhance drone performance, autonomy, and operational range, making them more viable for widespread use (Li et al., 2023).
- Regulatory Progress: Evolving regulatory frameworks will facilitate the safe integration of drones into airspaces, enabling more extensive and efficient operations (Turgut & Şeker, 2023).
- Business Models: Emerging business models such as drone-as-a-service (DaaS) and hybrid delivery systems will lower entry barriers and optimize logistics processes, providing flexible and cost-effective solutions (Benarbia & Kyamakya, 2021).

4.11.3 Final Thoughts on the Future of Drone Technology in Logistics

With the ability to significantly alter supply chain dynamics, improve global logistics efficiency, and support sustainable economic growth, drone technology appears to have a bright future in logistics. As technology advances and regulatory landscapes evolve, drones will become an integral part of modern logistics, offering innovative solutions to complex logistical challenges.

The adoption of drones in logistics is set to revolutionize the industry, providing faster, cheaper, and more environmentally friendly delivery options. Continued investment in research, technology development, and regulatory collaboration will be essential to realize the full potential of drone logistics.

REFERENCES

Benarbia, T. & Kyamakya, K. (2021), 'A literature review of drone-based package delivery logistics systems and their implementation feasibility', *Sustainability*. https://doi.org/10.3390/su14010360

Bhatt, K., Pourmand, A. & Sikka, N. (2018), 'Targeted applications of unmanned aerial vehicles (drones) in telemedicine', *Telemedicine Journal and e-Health* 24(11), 833–838. https://doi.org/10.1089/tmj.2017.0289

Braun, J., Gertz, S., Furer, A., Bader, T., Frenkel, H., Chen, J., Glassberg, E. & Nachman, D. (2019), 'The promising future of drones in prehospital medical care and its application to battlefield medicine', *Journal of Trauma and Acute Care Surgery*. https://doi.org/10.1097/TA.0000000000002221

Chiang, W., Li, Y., Shang, J. & Urban, T. (2019), 'Impact of drone delivery on sustainability and cost: Realizing the UAV potential through vehicle routing optimization', *Applied Energy*. https://doi.org/10.1016/J.APENERGY.2019.03.117

Das, D., Sewani, R., Wang, J. & Tiwari, M. (2021), 'Synchronized truck and drone routing in package delivery logistics', *IEEE Transactions on Intelligent Transportation Systems* 22, 5772–5782. https://doi.org/10.1109/tits.2020.2992549

Enarbia, T. & Kyamakya, K. (2021), 'A literature review of drone-based package delivery logistics systems and their implementation feasibility', *Sustainability*. https://doi.org/10.3390/su14010360

Euchi, J. (2020), 'Do drones have a realistic place in a pandemic fight for delivering medical supplies in healthcare systems problems?', *Chinese Journal of Aeronautics*. https://doi.org/10.1016/j.cja.2020.06.006

Godbole, G., Bhadekar, K., Singh, S. & Aranke, S. (2019), 'Design and development of a fixed wing un-manned aerial vehicle (UAV) prototype for crop spraying', *Journal of Emerging Technologies and Innovative Research*.

Gunaratne, K., Thibbotuwawa, A., Vasegaard, A., Nielsen, P. & Perera, H. (2022), 'Unmanned aerial vehicle adaptation to facilitate healthcare supply chains in low-income countries', *Drones*. https://doi.org/10.3390/drones6110321

Hassanalian, M. & Abdelkefi, A. (2017), 'Classifications, applications, and design challenges of drones: A review', *Progress in Aerospace Sciences* 91, 99–131. https://doi.org/10.1016/J.PAEROSCI.2017.04.003

Korkmaz, K. & Mohammed, S. (2020), 'Evaluating the use of drones in the area of transporta-tion/construction'. https://doi.org/10.30564/jaeser.v3i1.1787

Law, C., Moenig, C., Jeilani, H., Jeilani, M. & Young, T. (2023), 'Transforming healthcare logistics and evaluating current use cases of UAVs (drones) as a method of transportation in healthcare to generate recommendations for the NHS to use drone technology at scale: A narrative review', *BMJ Innovations* 9, 150–164. https://doi.org/10.1136/bmjinnov-2021-000861

Li, X., Tupayachi, J., Sharmin, A. & Ferguson, M. (2023), 'Drone-aided delivery methods, challenge, and the future: A methodological review', *Drones*. https://doi.org/10.3390/drones7030191

Macrina, G., Pugliese, L., Guerriero, F. & Laporte, G. (2020), 'Drone-aided routing: A literature review', *Transportation Research Part C: Emerging Technologies* 120, 102762. https://doi.org/10.1016/j.trc.2020.102762

Moshref-Javadi, M. & Winkenbach, M. (2021), 'Applications and research avenues for drone-based models in logistics: A classification and review', *Expert Systems with Applications* 177, 114854. https://doi.org/10.1016/J.ESWA.2021.114854

Nurgaliev, I., Eskander, Y. & Lis, K. (2023), 'The use of drones and autonomous vehicles in logistics and delivery', *Logistics and Transport.* https://doi.org/10.26411/83-1734-2015-2-55-6-23

Nyaaba, A. & Ayamga, M. (2021), 'Intricacies of medical drones in healthcare delivery: Implications for Africa', *Technology in Society.*

Oigbochie, A., Odigie, E. & Adejumo, B. (2021), 'Importance of drones in healthcare delivery amid a pandemic: Current and generation next application', *Open Journal of Medical Research* 2(1). https://doi.org/10.52417/OJMR.V2I1.187

Park, J., Kim, S. & Suh, K. (2018), 'A comparative analysis of the environmental benefits of drone-based delivery services in urban and rural areas', *Sustainability* 10, 888. https://doi.org/10.3390/SU10030888

Peng, K., Du, J., Lu, F., Sun, Q., Dong, Y., Zhou, P. & Hu, M. (2019), 'A hybrid genetic algorithm on routing and scheduling for vehicle-assisted multi-drone parcel delivery', *IEEE Access* 7, 49191–49200. https://doi.org/10.1109/ACCESS.2019.2910134

Raj, A. & Sah, B. (2019), 'Analyzing critical success factors for implementation of drones in the logistics sector using grey-dematel based approach', *Computers & Industrial Engineering* 138. https://doi.org/10.1016/j.cie.2019.106118

Rejeb, A., Rejeb, K., Simske, S. & Treiblmaier, H. (2021), 'Drones for supply chain management and logistics: A review and research agenda', *International Journal of Logistics Research and Applications* 26, 708–731. https://doi.org/10.1080/13675567.2021.1981273

Saunders, J., Saeedi, S. & Li, W. (2021), 'Autonomous aerial delivery vehicles, a survey of techniques on how aerial package delivery is achieved', *ArXiv.*

Stolaroff, J., Samaras, C., O'Neill, E., Lubers, A., Mitchell, A. & Ceperley, D. (2018), 'Energy use and life cycle greenhouse gas emissions of drones for commercial package delivery', *Nature Communications* 9. https://doi.org/10.1038/s41467-017-02411-5

Thomaidis, N. & Zeimpekis, V. (2023), Investigation of operational parameters that affect the use of drones in goods' stock count process: Evidence from experimental results', *International Joint Conference on Industrial Engineering and Operations Management Proceedings.* https://doi.org/10.14488/ijcieom2023$_a$bst$_0$042$_3$7772

Turgut, M. & Şeker, B. (2023), 'Drone technology in transportation management: A systematic review and framework for future research', *Journal of Aviation.* https://doi.org/10.30518/jav.1277694

5 Utilizing AI and UAVs for Disaster Response in Smart Cities

Bhupinder Singh and Christian Kaunert

5.1 INTRODUCTION AND BACKGROUND

Natural disasters have the potential to cause enormous socioeconomic losses and harm. The real damage and costs resulting from these threats have been increasing over the past few decades. As such, there is a growing need on the part of disaster managers to protect their communities in advance by creating effective management plans (Abbas et al., 2023). Artificial intelligence (AI) techniques are used in many research projects to evaluate data connected to disasters, which helps with well-informed disaster management. The UAVs that are powered by AI analyze dynamic variables like traffic, population density, and terrain mapping to optimize resource allocation (Ariyachandra et al., 2023). From dynamically adjusting flight paths to providing instant data analysis, these systems empower emergency responders with actionable insights, ensuring a more agile and informed crisis response. UAVs equipped with AI sensors provide real-time surveillance and data fusion capabilities. Integrating information from various sources, including thermal imaging and environmental sensors, these UAVs deliver a complete situational awareness, a critical component for meticulous disaster management. This automation ensures swift logistics, allowing for the timely deployment of resources during and after a disaster. AI-powered UAVs serve as communication hubs during crises (Samarakkody et al., 2023). Equipped with advanced communication technologies, including emergency broadcasting and mobile network support, these platforms enhance coordination among emergency response teams and provide real-time updates to affected communities. AI-powered UAVs conduct rapid damage assessments using image recognition and analysis (Van Hoang et al., 2024). This quick and accurate assessment informs immediate response strategies, allowing for more efficient allocation of resources and a faster recovery process. AI-generated insights from UAV data contribute to post-disaster recovery planning. A thorough analysis of infrastructure damage, economic impact, and community needs can help city planners formulate resilient recovery plans that prioritize both immediate necessities and long-term sustainability (Ullah et al., 2020).

 DOI: 10.1201/9781003481584-5

The specialty field of environmental health risk management is concerned with anticipating, organizing, responding to, and recovering from calamities (Alqurashi et al., 2022, Dias Santana et al., 2021). The main objectives of preserving lives, reducing suffering, and guaranteeing survivors receive the best medical treatment possible involve providing medical care both during and after such occurrences (Ahmed et al., 2022). In addition to man-made damages like industrial accidents and terrorist attacks, environmental health risk management is essential for managing natural catastrophes like hurricanes, earthquakes, and floods as well as public health situations like pandemics (Damaševičius et al., 2023, Singh, 2024).

5.1.1 Significance of Study

The growing frequency of worldwide disasters has caused a notable rise in the significance of environmental health risk management (Mahor et al., 2022). Artificial intelligence (AI)-driven technology and innovation have been included in environmental health risk management courses in recent years (Abid et al., 2023). Artificial Intelligence (AI) has shown to be a useful instrument in disaster management, providing improved capacities for precise forecasting and early warning systems (Singh, 2023).

AI combined with global luminescence can enhance flood prediction. With the use of artificial intelligence and data from the Global Navigation Satellite System, researchers have accurately forecast the amplitudes of tsunamis (Yang et al., 2020). AI technology also has the ability to provide early earthquake warnings. AI and machine learning may also be used efficiently in storm forecasting (Awan et al., 2021, Al-Turjman, 2020). There are serious humanitarian and economic difficulties as a result of the plague of terrorism (Khan et al., 2021). Artificial intelligence (AI) has been used in research projects to forecast and thwart terrorist operations, providing predictive capabilities and supporting preventative efforts (Singh, 2023, Khan et al., 2021). AI and machine learning together provide a chance to anticipate and stop impending terrorist acts. Artificial Intelligence becomes an invaluable tool in public health crises. It greatly aids in fast decision-making and appropriate resource allocation, which is crucial for efficient crisis management (Prawiyogi et al., 2022).

The advancements in technology have improved our ability to anticipate, respond to, and recover from these calamities. The terms "Radars", "Satellite imaging", and "Remote Sensing" describe a variety of technologies that allow us to gather information on the Earth's atmosphere and surface from a distance (Yu & Zikria, 2020). Thanks to technical improvements, it is now possible to monitor natural disasters (NDs) in real time, such as wildfires, landslides, and volcanic eruptions (Tanwar et al., 2019). Planning for evacuation and recovery can be made easier by analyzing data gathered by remote sensing technology which may help us better comprehend the scope and intensity of natural disasters (Zahmatkesh & Al-Turjman, 2020). Information may be gathered and shared quickly because of technologies like social media, cellphones, and the Internet of Things (IoT). IoT devices give first responders access to real-time data during NDs (Saleem et al., 2022). Examples of these devices are weather sensors and earthquake detectors. People in impacted areas can receive important information about evacuation routes, emergency shelters, and other assistance via social media and smartphone alerts (Sharma & Singh, 2022).

5.1.2 Increasing Frequency and Impact of Natural Disasters: Need for Innovative Solutions

Natural disasters include phenomena such as hurricanes, floods, tornadoes, landslides, tsunamis, earthquakes, volcanic eruptions, wildfires, droughts, and very high or low temperatures. These events can have alarming effects on human life as well as infrastructure (Ullah et al., 2020, Yan et al., 2020). Natural disasters have historically resulted in enormous devastation and a high death toll from events including storms, floods, wildfires, and volcanic eruptions (Alsamhi et al., 2021). The 1931 China floods, which covered an area of more than 50,000 square miles and killed an estimated 3.7 million people, were the deadliest natural disasters (NDs) in history. They were brought on by heavy rainfall and river overflows (Singh, 2022). The Krakatoa eruption in Indonesia in 1883, which killed over 36,000 people and caused a massive tsunami that impacted coastal communities throughout the region, is one of the other notable historical NDs. The 1985 Mexico City earthquake, which had an estimated magnitude of 8.0, killed over 10,000 people, left over 100,000 homeless, and seriously damaged the city's infrastructure. NDs have had and still have a big impact on communities all over the world (Xia et al., 2022). Over 15,000 people died and much of Japan's infrastructure was destroyed in the 2011 Tohoku earthquake and tsunami (Fatemidokht et al., 2021). Hurricane Maria struck Puerto Rico in 2017, wreaking havoc and knocking out electricity for months across most of the island. After natural disasters (NDs), communities frequently struggle to recover and reconstruct their life (Beneicke et al., 2019). They must deal with the damage of houses and

Year	Volcanic Activity	Wildfire	Extreme Temperature	Landslide	Drought	Earthquake	Storm	Flood
1900	1	0	0	0	0	1	0	0
1905	0	0	0	0	0	1	0	0
1910	0	0	0	0	0	2	0	0
1915	0	0	0	0	0	2	1	0
1920	0	0	0	0	0	2	1	0
1925	1	0	0	0	0	3	1	1
1930	0	0	0	0	0	4	2	2
1935	0	0	0	0	0	4	2	3
1940	0	0	0	0	0	5	2	5
1945	1	0	1	1	0	5	3	6
1950	2	1	2	1	0	6	4	7
1955	3	2	2	1	0	7	5	8
1960	4	3	3	2	0	9	7	10
1965	5	5	4	3	1	12	9	12
1970	6	7	5	4	1	14	11	15
1975	9	10	6	5	2	18	14	20
1980	12	14	8	7	3	22	19	26
1985	16	18	11	9	5	27	25	33
1990	18	22	15	12	8	33	32	42
1995	20	28	20	15	12	40	40	55
2000	25	35	28	20	18	50	52	70
2005	28	40	35	25	25	60	65	90
2010	35	50	40	30	35	75	85	120
2015	40	60	50	38	45	90	100	150
2020	44	75	60	50	55	110	130	180

FIGURE 5.1 Trend of global natural disaster occurrence 1900–2022

(Source: NATURAL DISASTER DATABOOK 2022)

infrastructure, the uprooting of residents, and the degradation of the environment, with consequences that can continue for years or even decades. The frequency and intensity of NDs are increased by climate change, making it more important than ever to prepare for and mitigate their consequences (Singh, 2022). Even though the effects of natural disasters (NDs) have been lessened by technological advancements like early warning systems, satellite navigation, cloud computing, drone technology, predictive analytics, machine learning, monitoring, communication, information sharing, geographical information systems (GIS), early flood detection systems, and building and infrastructure resilience (Jain et al., 2021). To reduce the effect on people, property, and the environment, continued investment in disaster planning and response initiatives is crucial (Ulusar et al., 2020).

5.1.3 Role of Artificial Intelligence (AI) in Disaster Management

Disaster management, which is defined as an essential set of strategies and actions intended to deal with disasters, is critical in reducing negative effects by focusing on activities related to prevention, response and recovery. Protecting people and physical assets, like buildings and infrastructure, comes first (Rahman et al., 2023 - Singh, 2022). It includes organizing resources, putting emergency and resilience plans into action, and encouraging cooperation between various entities, like local communities, governments, and international organizations (Nguyen et al., 2021). The complexity of this situation calls for the instantaneous understanding, association, and handling of data from several sources in addition to the coordination of broad reaction plans and public education (Al Ridhawi et al., 2021).

The ability to respond promptly is hindered by a basic barrier, even with the quantity of accessible data and sensor technologies: the volume of data exceeds our ability to derive useful insights from it. Breaking over this obstacle calls for significant innovations and inventive work (Ahammed et al., 2023 - Chen et al., 2019). It is theoretically possible to manage emergency situations more effectively by utilizing

Disaster Type	1992-2021 (30-year ave.)	2022
Wildfire	12	15
Volcanic Activity	5	5
Storm	100	105
Flood	147	177
Landslide	18	17
Extreme Temperature	17	12
Earthquake	26	31
Drought	16	23

FIGURE 5.2 Global disaster occurrence by disaster type 1992–2021 vs 2022

(Source: NATURAL DISASTER DATABOOK 2022)

AI technologies and systems (Refaai et al., 2022). This is made possible by accurate real-time data analysis, machine learning, and the automation of crucial activities. AI has the ability to completely change the way it handles disaster management while acting as a powerful ally for both national and international security (Thakur et al., 2021).

5.2 EMERGENCIES, CRISES, AND DISASTERS (ECDS) LANDSCAPE

Natural and man-made destruction affect millions of people annually and can result in substantial loss of life and bodily harm (Matthew et al., 2021). These things frequently happen out of the blue, shocking, at times, regions of residence. People who go through such tragedies might be distressed emotionally, with symptoms like despair and anxiety, and often face trouble falling asleep (Singh, 2019 - Solanki & Solanki, 2020). While many people can heal with the help of their families and the community, other people might need extra care to manage and get closer to recovery. This includes first responders and recovery personnel who are also at danger, as well as survivors in the impacted areas. Large-scale geological or meteorological occurrences that have the potential to cause property or human casualties are categorized as natural disasters (Srihith et al., 2022).

Tornadoes, hurricanes, strong storms, floods, wildfires, earthquakes, and droughts are a few examples. Reports of severe storms and floods are common in the US, necessitating state and municipal preparation for public asset protection and evacuation. Support is offered to people in need prior to, during, and following natural disasters through the Disaster Distress Helpline (Lima & Terán, 2019). Disasters created by humans, such as mass shootings, terrorism, industrial accidents, and shootings, can also seriously injure people and property. These traumatic occurrences,

Country	Disaster Occurrences (2022)
Indonesia	20
China	12
Philippines	12
Thailand	11
Afghanistan	8
Vietnam	8
Japan	7
India	7
Nepal	6
Malaysia	6

FIGURE 5.3 Top 10 countries in Asia with high occurrence of disasters in 2022

(Source: NATURAL DISASTER DATABOOK 2022)

like natural disasters, may force evacuations and put a burden on mental health services in the impacted areas (Liu et al., 2020).

5.3 ARTIFICIAL INTELLIGENCE FOR SMART DISASTER RESILIENCE

Natural calamities, which may include everything from hurricanes and earthquakes to wildfires and floods, have been wreaking havoc on our planet for ages. Numerous lives and pieces of property are lost in the face of such events, which frequently leave villages in ruins. Even if we are unable to completely control or forecast these events, we may lessen their effects by utilizing innovation and technology (Alsamhi et al., 2021). With its amazing potential, artificial intelligence (AI) is becoming a potent weapon in the fight against destruction by supporting preparedness, early warning systems, and response. Natural disasters can be roughly categorized as phenomena related to geology, meteorology, hydrology, and climate. Hurricanes, tornadoes, and blizzards are examples of meteorological adversities, whereas earthquakes, volcanic eruptions, and tsunamis are considered geological disasters. Destructions classified as hydrological include floods and landslides, whereas the rest are clubbed as climatological including heat waves, wildfires, and droughts (Janeera et al., 2021).

Artificial intelligence (AI) may be used to better understand natural catastrophes by using large and diverse information, such as geographical data, social media data, and wireless network sensor data. This includes natural disaster management (NDM) forecasting, detection, and humanitarian relief. Several communities are making significant investments in the development and/or use of AI technology for efficient natural disaster management as a result of realizing this potential (Heidari et al., 2023). Because AI can handle massive amounts of data, it may be able to lessen the effects of natural disasters. A higher level of productivity in distributing relief resources, mitigation shoots up the capability of mitigation actions and offers insightful information for anticipating natural disasters. AI and Machine Learning (ML) improve the judgment and actions of frontline relief workers while hastening the delivery of help to impacted individuals and locations (Durán et al., 2019). AI can help in multifarious ways such as:

Damage Assessment: To quickly assess and classify building and infrastructure damage in disaster zones, it uses satellite images and machine learning algorithms through the xView2 visual computing project. Semantic segmentation in xView2 reduces the evaluation time from weeks to hours or minutes, enabling rescue teams and control centers to act quickly (Gohari et al., 2022).

AI-Assisted Communication: Through well-known social media platforms, AI-driven chatbots provide crucial information to individuals affected by disasters. By utilizing Natural Language Processing (NLP), these chatbots analyze vast quantities of linguistic data to swiftly and economically engage with a wide range of people and provide support to those who want it most (Kurunathan et al., 2023).

- **Artificial Intelligence in Social Media**: Following an event of natural destruction, social media posts can include vital information, but handling the vast amount of data can be difficult for humans (Khan et al., 2023). The AI Digital Response (AIDR) software provides disaster response teams with real-time processed data by automatically gathering and categorizing tweets. Organizations using AIDR can find humanitarian information to use in their disaster response efforts (Mohsan et al., 2023).
- **Cloud-Based Emergency Management**: Emergency response personnel can now command, direct, and monitor crises in real time from any device thanks to cloud-based AI-powered solutions (Hamadi, 2023). During a natural mishap, this platform makes it easier to manage devices, apps, and communication apps virtually. It also helps with alerting, emergency evacuation planning, and identifying people and groups that may be at risk (Alhammadi et al., 2024).
- **Predictive Analysis**: AI enables scientists to forecast hurricanes and earthquakes by analyzing previous meteorological data and seismic activity. Geoscope uses temperature and rainfall data analysis to forecast the possibility of flooding. Through the processing of particular regional data, AI also helps in the prediction of economic and human repercussions (Bayomi & Fernandez, 2023).
- **AI-Powered Dispatching**: AI valuably handles a large number of distress calls during natural calamities. The algorithms are used to improve rescue operations, and this results in faster emergency services and more accurate information being provided by contact centers (Zafeiropoulos et al., 2023).

5.4 CHALLENGES AND COMPLEXITIES

Predicting natural crises is a difficult task for AI, especially when it comes to data accessibility and quality. For AI algorithms to be effective, large amounts of high-quality data are necessary for precise predictions, but trustworthy data are hard to get in places like poor nations. When important events are included in the training dataset, AI-based techniques show promise; however, the scarcity of these events restricts the amount of pertinent data that can be obtained (Fadhel et al., 2024). The further obstacle is the limited use of AI. Private sector-led AI initiatives could work with governmental or non-governmental organizations to target certain populations, sometimes ignoring wider areas or low-income families. This increases the possibility of dispersed goals and advantages, which might result in companies integrating or maintaining the tools less effectively. The intricacy of natural calamities presents additional difficulties. There are several unidentified elements that impact these occurrences. Although AI may make insightful forecasts, it cannot take the place of human judgment and experience (Damodaran et al., 2024). There are different concerns of using AI in disaster management such as:

Data Privacy and Security: AI heavily depends on data, and this holds true for disaster prevention systems. Acquiring data from diverse sources, including personal devices and sensors, is essential for creating predictive models. However, this brings up significant concerns regarding data privacy and security (Mohapatra et al., 2023).

Privacy: When gathering data from individuals, it is imperative to ensure the protection of their privacy. AI models need to be crafted to anonymize and aggregate data, preventing the identification of individuals. Transparent data usage policies and informed consent play a vital role (Al Fouri & Sakher, 2023).

Security: As the reliance on data for disaster prediction grows, the risk of cyberattacks also increases. Safeguarding the data and systems from malicious actors becomes a critical concern, as an attack on these systems could lead to false warnings or other misinformation, posing potentially disastrous consequences (Mishra & Singh, 2023).

AI Bias: When algorithms are taught on data that contains prejudices, they may inadvertently absorb such biases. Such biases may result in unfair resource prioritizing or erroneous forecasts in the field of adverse crisis prevention (Hassebo & Tealab, 2023). The AI model could not provide fair protection to everyone if historical data shows biases associated with particular locations or populations. In order to bring about a change with respect to the issue of bias in AI, careful data preparation and selection must be combined with ongoing model monitoring and modification. It should be included in ethical standards for AI development to identify and reduce prejudice (Bibri & Jagatheesaperumal, 2023).

Equity and Accessibility: It is crucial to make sure that everyone, regardless of socioeconomic background or geography, can use AI-driven solutions for shock avoidance. Communities that have historically been marginalized may have less access to resources and technology, making them more susceptible to unfavorable natural occurrences (Chen et al., 2023). Encouraging fairness means figuring out the underlying inequalities in infrastructure, education, and resources that can make the effects of a crisis worse on vulnerable groups in addition to providing access to AI technologies (Ullah et al., 2024).

Over-reliance on Technology: Although artificial intelligence (AI) is a powerful tool, there is a chance that technology will be used excessively to forecast and lessen tragedies. Disaster response operations should continue to be centered on human judgment and experience (Ghaffarian et al., 2023). A dependence on AI that is too great might lead to complacency, when people blindly trust the technology and miss important details or warning indicators. Finding a balance between human expertise and AI-assisted decision-making is crucial. This is to make sure that AI enhances rather than replaces the jobs of emergency responders and disaster management specialists (Ouaissa et al., 2024).

Decision-Making: Clear lines of responsibility are essential as AI technologies are increasingly included in emergency management systems. An AI system needs to make explicit the reasoning behind its judgments when it makes forecasts or suggestions (AbdelAziz et al., 2024). There must be procedures in place for challenging AI-generated choices. Structures for monitoring and accountability are necessary to guarantee the moral and responsible application of AI in situations that impact human safety and life (Seong & Jiao, 2023).

5.5 LEVERAGING UNMANNED AERIAL VEHICLES (UAVS)

Unmanned Aerial Vehicles (UAVs) have become more and more important in disaster management since they can quickly analyze circumstances from the air, saving lives from danger. They may perform a wide range of activities, from damage assessment to search and rescue missions, because of their adaptability and array of sensors, cameras, and even payload delivery systems (Pandya et al., 2023). The introduction of Unmanned Aerial Vehicles (UAVs), sometimes known as drones, has revolutionized a number of industries, including disaster management. Drones provide a unique aerial viewpoint, improved mobility, and simple deployment, which make them invaluable resources for risk assessment and response. The use of UAVs in crisis management includes studying real-life cases and learning important lessons from them (Farazmehr & Wu, 2023).

During a natural calamity, disaster management professionals' quick action is critical to preserving lives in the impacted areas. The most effective method of situational awareness is aerial evaluation using UAV networks (Farazmehr & Wu, 2023).

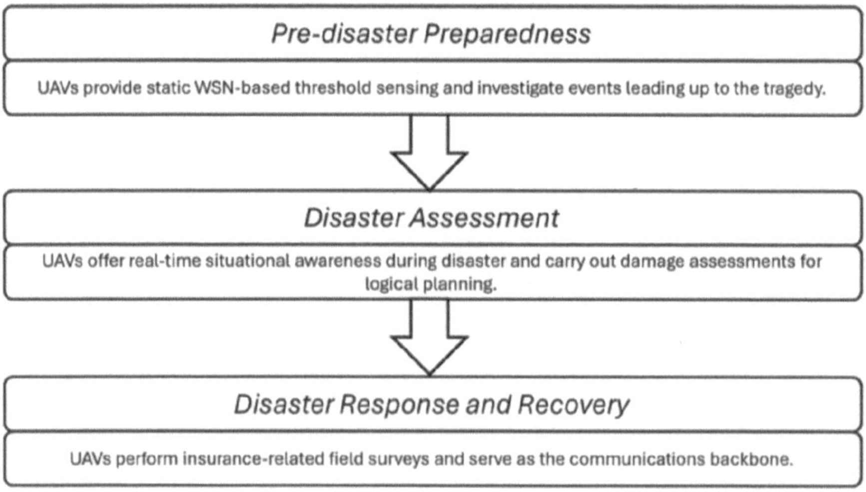

FIGURE 5.4 Usability of UAVs in disasters

(Source: Original)

Each country has its own regulations regarding the use of UAVs; however, during emergencies special permissions are usually given in order to enable first responders to quickly examine the situation. Using unmanned aerial vehicles (UAVs), first responders may analyze the impact of an incident on structures, determine the level of damage, examine the state of transportation infrastructure, and estimate the number of people who may be affected (Fakhraian et al., 2023).

The effectiveness of the UAV network can be hampered by issues including restricted power supplies, processing power restrictions, unstable communication channels, unplanned node failures, maximum physical load sizes, and mobility in challenging environments (Krichen & Mihoub, 2023). It is impossible to overestimate the significance of real-time information in guiding Search and Rescue (SAR) operations, and a recent Red Cross assessment identifies Unmanned Aerial Vehicles (UAVs) as one of the most potent and promising technologies for this use (Teh & Rana, 2023 - Wei et al., 2024). Teams of unmanned aerial vehicles (UAVs) can provide reconnaissance and mapping assistance and structural evaluations, locate stranded survivors and direct them to safe havens, and act as a mobile device ad hoc communications infrastructure by tying in with the closest radio access network (RAN). All of these functions can be carried out from higher ground. For these jobs, a variety of unmanned aerial vehicles (UAVs) can be utilized, such as balloons, blimps, and fixed-wing and rotary-wing models (Dou et al., 2023).

5.5.1 UAV Applications in Management Mechanism and Response-Resilience

Unmanned Aerial Vehicles (UAVs) have been shown useful in several disaster management domains, predominantly in the following areas such as:

Monitoring, Forecasting, and Early Alerts: UAVs may function as early warning systems by using structural and environmental monitoring as well as data analysis for predictions. Notifying populations that are susceptible to disasters in a timely manner is a crucial part of disaster prevention (Banafaa et al., 2024). AI-enabled systems are able to evaluate data from many sources such as social media, weather sensors, and satellites, to find early warning signs of approaching calamities. AI systems can evaluate atmospheric data to precisely forecast the path and severity of storms. These estimates enable authorities to warn people in advance and evacuate high-risk locations, perhaps saving many lives (Ali et al., 2023).

Information Fusion and Sharing: UAVs can help assist other applications during disaster management by fusing information from various sources or serving as a bridge between various information technologies (Iqbal et al., 2023). With the help of AI, knowledge and ability to anticipate earthquakes, another terrible natural disaster, have greatly increased. In order to predict seismic occurrences, machine learning algorithms may examine past seismic data, track ground motions, and identify minute alterations in the Earth's crust. While it may not be possible to completely prevent earthquakes, early detection gives individuals precious seconds or even minutes to seek cover which reduces the number of casualties (Pillai et al., 2024).

Standalone Communication System: In the event of a disaster, UAVs can be used to repair or replace communication infrastructure that has been damaged or destroyed. UAVs are useful for finding and rescuing people who are hurt, missing, or buried under rubble (Abid et al., 2023).

Preventing Forest Fires: AI-powered systems can be crucial in averting wildfires which are becoming more frequent and intense as a result of climate change (Heidari et al., 2023 - Rashid et al., 2023). Artificial intelligence (AI)-enabled drones can search forests for potential ignition sources, such as lightning strikes or bonfires. By evaluating climatic data, AI can also predict how a fire would spread, which aids firefighters in more effectively planning their activities (Singh et al., 2023).

Damage Assessment: UAVs use a variety of techniques such as UAV video inspection and structural health monitoring to assist in assessing damage. UAVs help with information collecting during the crisis phase, especially with regard to the movement of impacted individuals and the deployment of rescue teams. They also assist with situational awareness, logistics, and evacuation support (Singh et al., 2024).

Forecasting and Handling of Floods: With AI models, floods, a recurrent hazard that affects many locations worldwide, can be better predicted. In order to forecast when and where floods are likely to occur, these models process data from soil moisture sensors, river levels, and rainfall gauges. AI-driven flood modeling helps to reduce flood risk and damage by building better infrastructure and urban planning (Kumar et al., 2024). By utilizing chatbots, virtual assistants, and automated systems to improve communication between emergency responders, impacted communities, and government organizations, artificial intelligence (AI) improves the coordination of disaster response activities. Depending on the severity of the crisis, real-time data analysis allows for the effective deployment of resources (Singh et al., 2022).

5.5.2 Case Studies of UAVs in Disaster Management

Nepal Earthquake, 2015: The notable instances of utilizing UAVs in disaster management occurred in the aftermath of the destructive earthquake in Nepal in 2015. UAVs played a crucial role in swiftly assessing the damage in locations that were difficult for rescue teams to access (Singh & Park, 2022). They supplied high-resolution imagery that aided in creating maps depicting the severity of the destruction and determining safe paths for ground teams. This case underscored the UAVs' capacity to rapidly collect essential data, proving invaluable in the initial hours following a disaster (Singh et al., 2023).

2017's Hurricane Harvey: During Hurricane Harvey in the United States, unmanned aerial vehicles (UAVs) were crucial in tracking water levels, assessing damage to infrastructure, and supporting rescue efforts. They provided emergency responders with real-time data that was essential for them to identify regions that needed to be addressed right away. This case demonstrated how useful unmanned aerial vehicles (UAVs) may be in large-scale natural disaster situations, when accurate and timely information can have a big impact on relief operations (Singh et al., 2022).

2018 California Wildfires: During the California wildfires, unmanned aerial vehicles (UAVs) were widely used to track the spread of the fire, locate hotspots, and evaluate the damage to natural resources and property (Singh et al., 2021). They enabled firemen to come up with more potent plans and reactions, lowering dangers and making the most use of available resources. This incident demonstrated the value of unmanned aerial vehicles (UAVs) in providing continuous surveillance and useful intelligence in ever-changing natural calamity situations (Singh et al., 2022).

Indian States Floods, 2018–2022: Following the massive floods that struck Kerala in 2018, Uttarakhand in 2021, and Assam in 2022, drones were utilized to identify persons who were reported missing and assess the extent of the damage. Rescue crews were able to plan their operations more effectively and save more lives because of the real-time information provided by the drones (Kumar et al., 2023).

5.5.3 Use of UAVs: Lessons Learned

Need for Quick Deployment and Flexibility: A crucial lesson highlights how important quick deployment is. Time is of the essence in crisis situations, and UAVs' ability to be quickly deployed can have a big influence on how

FIGURE 5.5 Case studies

(Source: Original)

things turn out. UAVs' flexibility in navigating a variety of situations like navigating through debris left over from earthquakes or keeping an eye on large regions damaged by floods is essential (Kumar et al., 2023).

Data Analysis and Integration: UAVs are useful not just for gathering data but also for integrating and analyzing that data. Making informed decisions in each circumstance requires the ability to swiftly examine and interpret aerial data. This emphasizes the need for strong data processing systems and trained staff members who are adept at analyzing real-time UAV data (Singh et al., 2021).

Ethics and Regulation Concerns: Regulation and morality are brought up by the use of UAVs. It is essential to make sure that UAV activities comply with national and international legislation, especially in situations involving cross-border disasters. It is important to pay close attention to privacy and data security problems, particularly when they include civilian populations (Singh et al., 2021).

Education and Readiness: Personnel with proper training is essential for the successful use of UAVs in disaster management. This includes not just UAV operators but also decision-makers who analyze and utilize the data that is supplied. Frequent preparation and training sessions greatly increase the effectiveness of UAV deployments in real-world natural calamity management scenarios (Choi et al., 2022).

5.6 SMART CITIES AND RESILIENCE

Smart cities promote social cohesiveness and innovation deployment by integrating physical and IT systems and infrastructure (Singh et al., 2022). Resilience, or the ability to thrive in the face of challenges like population growth, climate change, and globalization, is a concept that is emerging for cities (Hossin et al., 2023). Cities that are resilient are able to withstand shocks, continue with regular operations in the face of disruptive occurrences, and exhibit proficiency in planning, absorbing, recovering, and adapting in several domains such as physical, cognitive, and social. In addition to classic preventative and hardening notions, resilience is seen as a proactive strategy that strives to strengthen infrastructure capacity to avoid damage, relieve suffering, and facilitate recovery (Park et al., 2022). An estimated 2.5 billion more people are anticipated to live in cities by 2050. And the likelihood and severity of such catastrophes are increased in metropolitan regions due to the concentration of people and resources. Sixty-eight percent of the world's population is expected to reside in urban areas by 2050. Infrastructure and application investments are made in cities to improve citizen operations, services, and the urban experience as a whole. Citizens can depend on uninterrupted basic services like public transit, communications, electricity and water distribution, hospitals, and schools when they live in a safe, resilient smart city equipped with critical infrastructure (Hazmy et al., 2023).

Gas stations, power plants, hospitals, banking and financial services, transportation, government buildings, military installations, water reservoirs, and bridges are examples of critical infrastructure (JayaLakshmi et al., 2023 - Waghmare et al., 2023). The operational continuity of vital infrastructure in smart cities can be disrupted by both natural and man-made hazards, such as cyber and physical assaults. There is a growing trend of combined attacks that have cascade effects on linked systems (Van Hoang, 2024). It is now difficult to protect cities against interruptions to essential infrastructure operations; it requires a variety of interdisciplinary approaches, a wide range of talents, and the capacity to evaluate large volumes of data from several sources, including social media. It is important to recognize the interdependencies across various categories of vital infrastructure and devise strategies to mitigate the negative ripple effects in order to guarantee the prompt restoration of service performance levels following disturbances (Alhammadi et al., 2024).

Developing a smart city that is resilient requires a dual strategy. First, it is critical to identify and evaluate possible risks and weak points, such as hurricanes, wildfires, floods, earthquakes, and cyberattacks. Mapping and tracking of temporal and geographical risk patterns is made possible by utilizing technologies such as big data analytics, remote sensing, and geographic information systems (Ullah et al., 2024 - Wei et al., 2024). This thorough assessment helps prioritize adaptation and mitigation strategies, enabling the best possible use of available resources. The capabilities for emergency response and recovery are the focus of the second crucial phase. Resilient smart cities minimize mortality and damage by responding to and recovering from disasters with speed and efficiency (Damodaran et al., 2024). Real-time monitoring and sharing of information about infrastructure and population status is made possible using sensors, cameras, drones, and smart grids. By clubbing blockchain, artificial intelligence, and cloud computing, the path for effective allocation of resources is laid, especially during emergency operations and relief efforts, enables data and system backup, and automates decision-making processes. Smart cities may strengthen their resilience to negative disasters by implementing these methods (AbdelAziz et al., 2024).

5.7 CONCLUSION AND FUTURE SCOPE

Artificial Intelligence has great potential to improve the world's resistance to disasters. AI has the potential to close gaps and improve results at several points before, during, and following a natural adversity, even if it cannot completely replace current disaster management methods. Encouraging stronger collaboration across disaster management organizations, guaranteeing increased availability of high-quality data, and using an agile strategy that incorporates machine and human intelligence may all contribute to improving the efficacy of disaster responses. Artificial Intelligence is swiftly revolutionizing our approach to disaster prevention and management. Its capacity to swiftly process and analyze extensive datasets in real time empowers us to predict, prepare for, and respond to natural disasters with greater efficacy.

Unmanned Aerial Vehicles (UAVs) have become revolutionary tools in disaster management, offering quick, flexible, and safe ways to evaluate and address different types of disasters. The case studies from the US and the Indian States provide verifiable evidence of their value. But to fully realize their potential, prompt implementation, data integration, regulatory compliance, and continuous training are important considerations. In the future, the knowledge gained from earlier deployments should guide the creation of more sophisticated UAV tactics and technology, enhancing their ability to manage emergencies and risks more successfully. Undoubtedly, more integration of UAVs will be necessary in the future of disaster management to improve reaction times, safety, and general efficiency.

As it persists in refining and advancing AI-powered tools and systems, we draw nearer to a future where we can significantly mitigate the devastating impact of these events on our communities and the planet. However, it is crucial to approach AI in disaster prevention with caution, addressing ethical concerns and ensuring fair access to these life-saving technologies. By doing so, it can fully leverage the potential of AI to shield ourselves and future generations from the unpredictable forces of nature.

REFERENCES

Abbas, N., Abbas, Z., Liu, X., Khan, S. S., Foster, E. D., & Larkin, S. (2023). A survey: Future smart cities based on advance control of Unmanned Aerial Vehicles (UAVs). *Applied Sciences*, *13*(17), 9881.

AbdelAziz, N. M., Eldrandaly, K. A., Al-Saeed, S., Gamal, A., & Abdel-Basset, M. (2024). Application of GIS and IoT Technology based MCDM for Disaster Risk Management: Methods and case study. *Decision Making: Applications in Management and Engineering*, *7*(1), 1–36.

Abid, S. K., Chan, S. W., Sulaiman, N., Bhatti, U., & Nazir, U. (2023, October). Present and future of artificial intelligence in disaster management. In *2023 International Conference on Engineering Management of Communication and Technology (EMCTECH)* (pp. 1–7). IEEE.

Ahammed, T. B., Patgiri, R., & Nayak, S. (2023). A vision on the artificial intelligence for 6G communication. *ICT Express*, *9*(2), 197–210.

Ahmed, R., Chen, Y., & Hassan, B. (2022). Deep residual learning-based cognitive model for detection and classification of transmitted signal patterns in 5G smart city networks. *Digital Signal Processing*, *120*, 103290.

Al Fouri, A., & Sakher, S. (2023). Artificial intelligence on the smart city. *Remittances Review*, *8*(4), 4197–4210.

Alhammadi, A., Abraham, A., Fakhreddine, A., Tian, Y., Du, J., & Bader, F. (2024). Envisioning the future role of 3D wireless networks in preventing and managing disasters and emergency situations. *arXiv preprint arXiv:2402.10600.*

Ali, S. A., Elsaid, S. A., Ateya, A. A., ElAffendi, M., & El-Latif, A. A. A. (2023). Enabling technologies for next-generation smart cities: A comprehensive review and research directions. *Future Internet*, *15*(12), 398.

Alqurashi, F. A., Alsolami, F., Abdel-Khalek, S., Sayed Ali, E., & Saeed, R. A. (2022). Machine learning techniques in internet of UAVs for smart cities applications. *Journal of Intelligent & Fuzzy Systems*, *42*(4), 3203–3226.

Al Ridhawi, I., Bouachir, O., Aloqaily, M., & Boukerche, A. (2021). Design guidelines for cooperative UAV-supported services and applications. *ACM Computing Surveys (CSUR)*, *54*(9), 1–35.

Alsamhi, S. H., Afghah, F., Sahal, R., Hawbani, A., Al-Qaness, M. A., Lee, B., & Guizani, M. (2021). Green internet of things using UAVs in B5G networks: A review of applications and strategies. *Ad Hoc Networks*, *117*, 102505.

Alsamhi, S. H., Almalki, F. A., Al-Dois, H., Othman, S. B., Hassan, J., Hawbani, A., & Saleh, H. (2021). Machine learning for smart environments in B5G networks: Connectivity and QoS. *Computational Intelligence and Neuroscience*, *2021*. https://doi.org/10.1155/2021/6805151

Al-Turjman, F. (Ed.). (2020). *Unmanned Aerial Vehicles in Smart Cities*. Springer Nature.

Ariyachandra, M. M. F., & Wedawatta, G. (2023). Digital twin smart cities for disaster risk management: A review of evolving concepts. *Sustainability*, *15*(15), 11910.

Awan, N., Khan, S., Rahmani, M. K. I., Tahir, M., Alam, N., Alturki, R., & Ullah, I. (2021). Machine learning-enabled power scheduling in IoT-based smart cities. *Computers, Materials and Continua*, *67*(2), 2449–2462.

Bayomi, N., & Fernandez, J. E. (2023). Eyes in the sky: Drones applications in the built environment under climate change challenges. *Drones*, *7*(10), 637.

Banafaa, M., Pepeoğlu, Ö., Shayea, I., Alhammadi, A., Shamsan, Z., Razaz, M. A., & Al-Sowayan, S. (2024). A comprehensive survey on 5G-and-beyond networks with UAVs: Applications, emerging technologies, regulatory aspects, research trends and challenges. *IEEE Access*.

Beneicke, J., Juan, A. A., Xhafa, F., Lopez-Lopez, D., & Freixes, A. (2019). Empowering citizens' cognition and decision making in smart sustainable cities. *IEEE Consumer Electronics Magazine*, *9*(1), 102–108.

Bibri, S. E., & Jagatheesaperumal, S. K. (2023). Harnessing the potential of the metaverse and artificial intelligence for the internet of city things: Cost-effective XReality and synergistic AIoT technologies. *Smart Cities*, *6*(5), 2397–2429.

Chen, L., Chen, Z., Zhang, Y., Liu, Y., Osman, A. I., Farghali, M., & Yap, P. S. (2023). Artificial intelligence-based solutions for climate change: A review. *Environmental Chemistry Letters*, 1–33. https://doi.org/10.1007/s10311-023-01617-y

Chen, M., Challita, U., Saad, W., Yin, C., & Debbah, M. (2019). Artificial neural networks-based machine learning for wireless networks: A tutorial. *IEEE Communications Surveys & Tutorials*, *21*(4), 3039–3071.

Choi, M., Azzaoui, A. E., Singh, S. K., Salim, M. M., Jeremiah, S. R., & Park, J. H. (2022). The future of metaverse: Security issues, requirements, and solutions. *Human-Centric Computing and Information Sciences*, *12*(60), 1–14.

Damaševičius, R., Bacanin, N., & Misra, S. (2023). From sensors to safety: Internet of Emergency Services (IoES) for emergency response and disaster management. *Journal of Sensor and Actuator Networks*, *12*(3), 41.

Damodaran, D., Kumar, D., Damodaran, S., & Gopalakrishnan, S. (2024). Futuristic disaster mitigation: The role of GPUs and AI accelerators. In *Predicting Natural Disasters with AI and Machine Learning* (pp. 173–201). IGI Global.

Dias Santana, G. M., Cristo, R. S. D., & Lucas Jaquie Castelo Branco, K. R. (2021). Integrating cognitive radio with unmanned aerial vehicles: An overview. *Sensors*, *21*(3), 830.

Dou, X., Chen, W., Zhu, L., Bai, Y., Li, Y., & Wu, X. (2023). Machine learning for smart cities: A comprehensive review of applications and opportunities. *International Journal of Advanced Computer Science and Applications*, *14*(9). https://doi.org/10.14569/ijacsa.2023.01409104

Durán, C. A., Córdova, F. M., & Palominos, F. (2019). A conceptual model for a cyber-social-technological-cognitive smart medium-size port. *Procedia Computer Science*, *162*, 94–101.

Fadhel, M. A., Duhaim, A. M., Saihood, A., Sewify, A., Al-Hamadani, M. N., Albahri, A. S., & Gu, Y. (2024). Comprehensive systematic review of information fusion methods in smart cities and urban environments. *Information Fusion*, *107*, 102317.

Fakhraian, E., Semanjski, I., Semanjski, S., & Aghezzaf, E. H. (2023). Towards safe and efficient unmanned aircraft system operations: literature review of digital twins' applications and european union regulatory compliance. *Drones*, *7*(7), 478.

Farazmehr, S., & Wu, Y. (2023). Locating and deploying essential goods and equipment in disasters using AI-enabled approaches: A systematic literature review. *Progress in Disaster Science*, *19*, 100292.

Fatemidokht, H., Rafsanjani, M. K., Gupta, B. B., & Hsu, C. H. (2021). Efficient and secure routing protocol based on artificial intelligence algorithms with UAV-assisted for vehicular ad hoc networks in intelligent transportation systems. *IEEE Transactions on Intelligent Transportation Systems*, *22*(7), 4757–4769.

Ghaffarian, S., Taghikhah, F. R., & Maier, H. R. (2023). Explainable artificial intelligence in disaster risk management: Achievements and prospective futures. *International Journal of Disaster Risk Reduction*, *98*, 104123.

Gohari, A., Ahmad, A. B., Rahim, R. B. A., Supa'at, A. S. M., Abd Razak, S., & Gismalla, M. S. M. (2022). Involvement of surveillance drones in smart cities: A systematic review. *IEEE Access*, *10*, 56611–56628.

Hamadi, R. (2023). *Artificial Intelligence Applications in Intrusion Detection Systems for Unmanned Aerial Vehicles* (Doctoral dissertation).

Hassebo, A., & Tealab, M. (2023). Global models of smart cities and potential IoT applications: A review. *IoT*, *4*(3), 366–411.

Hazmy, A. I., Hawbani, A., Wang, X., Al-Dubai, A., Ghannami, A., Yahya, A. A., & Alsamhi, S. H. (2023). Potential of satellite-airborne sensing technologies for agriculture 4.0 and climate-resilient: A review. *IEEE Sensors Journal*, *24*(4), 4161–4180.

Heidari, A., Jafari Navimipour, N., Unal, M., & Zhang, G. (2023). Machine learning applications in internet-of-drones: systematic review, recent deployments, and open issues. *ACM Computing Surveys*, *55*(12), 1–45.

Hossin, M. A., Chen, L., Asante, I. O., Boadi, E. A., & Adu-Yeboah, S. S. (2023). Climate change and COP26: Role of information technologies in disaster management and resilience. *Environment, Development and Sustainability*, *27*(3), 5659–5685.

Iqbal, M., Ghani, M., Tabrez, S., & Mehsud, A. K. (2023). Scope of artificial intelligence in enhancement of emergency rescue services: Future prospects. *Al-Qanṭara*, *9*(3).

Jain, R., Nagrath, P., Thakur, N., Saini, D., Sharma, N., & Hemanth, D. J. (2021). Towards a smarter surveillance solution: The convergence of smart city and energy efficient unmanned aerial vehicle technologies. *Development and Future of Internet of Drones (IoD): Insights, Trends and Road Ahead*, 109–140.

Janeera, D. A., Gnanamalar, S. S. R., Ramya, K. C., & Kumar, A. A. (2021). Internet of things and artificial intelligence-enabled secure autonomous vehicles for smart cities. In *Automotive Embedded Systems: Key Technologies, Innovations, and Applications* (pp. 201–218). Springer International Publishing.

Khan, I. U., Alturki, R., Alyamani, H. J., Ikram, M. A., Aziz, M. A., Hoang, V. T., & Cheema, T. A. (2021). RSSI-controlled long-range communication in secured IoT-enabled unmanned aerial vehicles. *Mobile Information Systems*, *2021*, 1–11.

Khan, S. M., Shafi, I., Butt, W. H., Diez, I. D. L. T., Flores, M. A. L., Galán, J. C., & Ashraf, I. (2023). A systematic review of disaster management systems: approaches, challenges, and future directions. *Land*, *12*(8), 1514.

Krichen, M., & Mihoub, A. (2023). Unmanned aerial vehicles communications security challenges: A survey. In *Unmanned Aerial Vehicles Applications: Challenges and Trends* (pp. 349–373). Springer International Publishing.

Kumar, M., Kim, C., Son, Y., Singh, S. K., & Kim, S. (2024). Empowering cyberattack identification in IoHT networks with neighborhood component-based improvised long short-term memory. *IEEE Internet of Things Journal*, *11*(9), 16638–16646.

Kumar, S., Singh, S. K., Varshney, S., Singh, S., Kumar, P., Kim, B. G., & Ra, I. H. (2023). Fusion of deep sort and Yolov5 for effective vehicle detection and tracking scheme in real-time traffic management sustainable system. *Sustainability*, *15*(24), 16869.

Kurunathan, H., Huang, H., Li, K., Ni, W., & Hossain, E. (2023). Machine learning-aided operations and communications of unmanned aerial vehicles: A contemporary survey. *IEEE Communications Surveys & Tutorials*. https://doi.org/10.1109/comst.2023.3312221

Lima, S., & Terán, L. (2019, April). Cognitive smart cities and deep learning: A classification framework. In *2019 Sixth International Conference on eDemocracy & eGovernment (ICEDEG)* (pp. 180–187). IEEE.

Liu, Y., Dai, H. N., Wang, Q., Shukla, M. K., & Imran, M. (2020). Unmanned aerial vehicle for internet of everything: Opportunities and challenges. *Computer Communications*, *155*, 66–83.

Mahor, V., Bijrothiya, S., Mishra, R., Rawat, R., & Soni, A. (2022). The smart city based on AI and infrastructure: A new mobility concepts and realities. *Autonomous Vehicles: Using Machine Intelligence*, *1*, 277–295.

Matthew, U. O., Kazaure, J. S., Onyebuchi, A., Daniel, O. O., Muhammed, I. H., & Okafor, N. U. (2021, February). Artificial intelligence autonomous unmanned aerial vehicle (UAV) system for remote sensing in security surveillance. In *2020 IEEE 2nd International Conference on Cyberspac (CYBER NIGERIA)* (pp. 1–10). IEEE.

Mishra, P., & Singh, G. (2023). Energy management systems in sustainable smart cities based on the internet of energy: A technical review. *Energies*, *16*(19), 6903.

Mohapatra, S., Sahoo, A., Mohanty, S., & Mohanty, S. N. (2023). Drone-enabled smart healthcare system for smart cities. *Drone Technology: Future Trends and Practical Applications*, 393–423. https://doi.org/10.1002/9781394168002.ch16

Mohsan, S. A. H., Othman, N. Q. H., Li, Y., Alsharif, M. H., & Khan, M. A. (2023). Unmanned aerial vehicles (UAVs): Practical aspects, applications, open challenges, security issues, and future trends. *Intelligent Service Robotics*, *16*(1), 109–137.

Nguyen, D. C., Ding, M., Pathirana, P. N., Seneviratne, A., Li, J., & Poor, H. V. (2021). Federated learning for internet of things: A comprehensive survey. *IEEE Communications Surveys & Tutorials*, *23*(3), 1622–1658.

Ouaissa, M., Ouaissa, M., Arezki, S., Ullah, I., Boulouard, Z., El Himer, S., & Iwendi, C. (2024, February). 2nd international workshop on intelligent systems for sustainable smart cities-proceedings of I3SC'23. In *2nd International Workshop on Intelligent Systems for Sustainable Smart Cities (I3SC'23)*. Unpublished.

Pandya, S., Srivastava, G., Jhaveri, R., Babu, M. R., Bhattacharya, S., Maddikunta, P. K. R., & Gadekallu, T. R. (2023). Federated learning for smart cities: A comprehensive survey. *Sustainable Energy Technologies and Assessments*, *55*, 102987.

Park, J. H., Singh, S. K., Salim, M. M., Azzaoui, A. E., & Park, J. H. (2022). Ransomware-based cyber attacks: A comprehensive survey. *Journal of Internet Technology*, *23*(7), 1557–1564.

Pillai, B. M., Suthakorn, J., Sivaraman, D., Nakdhamabhorn, S., Nillahoot, N., Ongwattanakul, S., & Magid, E. (2024). A heterogeneous robots collaboration for safety, security, and rescue robotics: e-ASIA joint research program for disaster risk and reduction management. *Advanced Robotics*, 1–23.

Prawiyogi, A. G., Purnama, S., & Meria, L. (2022). Smart cities using machine learning and intelligent applications. *International Transactions on Artificial Intelligence*, *1*(1), 102–116.

Rahman, K., Aziz, M. A., Usman, N., Kiren, T., Cheema, T. A., Shoukat, H., & Sajid, A. (2023). Cognitive lightweight logistic regression-based IDS for IoT-enabled FANET to detect cyberattacks. *Mobile Information Systems, 2023.*

Rashid, A. B., Kausik, A. K., Al Hassan Sunny, A., & Bappy, M. H. (2023). Artificial intelligence in the military: An overview of the capabilities, applications, and challenges. *International Journal of Intelligent Systems, 2023.*

Refaai, M. R. A., Dattu, V. S., Niranjana Murthy, H. S., Pramod Kumar, P., Kannadasan, B., & Diriba, A. (2022). An artificial intelligence mechanism for the prediction of signal strength in drones to IoT devices in smart cities. *Advances in Materials Science and Engineering, 2022.*

Saleem, M., Abbas, S., Ghazal, T. M., Khan, M. A., Sahawneh, N., & Ahmad, M. (2022). Smart cities: Fusion-based intelligent traffic congestion control system for vehicular networks using machine learning techniques. *Egyptian Informatics Journal*, *23*(3), 417–426.

Samarakkody, A., Amaratunga, D., & Haigh, R. (2023). Technological innovations for enhancing disaster resilience in smart cities: A comprehensive urban Scholar's analysis. *Sustainability*, *15*(15), 12036.

Seong, K., & Jiao, J. (2023). Is a smart city framework the key to disaster resilience? A systematic review. *Journal of Planning Literature*, 08854122231199462.

Sharma, A., & Singh, B. (2022). Measuring impact of e-commerce on small scale business: A systematic review. *Journal of Corporate Governance and International Business Law*, *5*(1).

Singh, B. (2019). Profiling public healthcare: A comparative analysis based on the multidimensional healthcare management and legal approach. *Indian Journal of Health and Medical Law*, *2*(2), 1–5.

Singh, B. (2022). COVID-19 pandemic and public healthcare: Endless downward spiral or solution via rapid legal and health services implementation with patient monitoring program. *Justice and Law Bulletin*, *1*(1), 1–7.

Singh, B. (2022). Relevance of agriculture-nutrition linkage for human healthcare: A conceptual legal framework of implication and pathways. *Justice and Law Bulletin*, *1*(1), 44–49.

Singh, B. (2022). Understanding legal frameworks concerning transgender healthcare in the age of dynamism. *Electronic Journal of Social and Strategic Studies*, *3*, 56–65.

Singh, B. (2023). Blockchain technology in renovating healthcare: Legal and future perspectives. In *Revolutionizing Healthcare Through Artificial Intelligence and Internet of Things Applications* (pp. 177–186). IGI Global.

Singh, B. (2024). Legal dynamics lensing metaverse crafted for videogame industry and E-sports: Phenomenological exploration catalyst complexity and future. *Journal of Intellectual Property Rights Law*, *7*(1), 8–14.

Singh, S. K., Azzaoui, A. E., Choo, K. K. R., Yang, L. T., & Park, J. H. (2023). A comprehensive survey on blockchain for secure IoT-enabled smart city beyond 5G: Approaches, processes, challenges, and opportunities. *Human-centric Computing and Information Sciences*, 13(51).

Singh, S. K., Kumar, M., Tanwar, S., & Park, J. H. (2024). GRU-based digital twin framework for data allocation and storage in IoT-enabled smart home networks. *Future Generation Computer Systems*, *153*, 391–402.

Singh, S. K., Lee, C., & Park, J. H. (2022). CoVAC: A P2P smart contract-based intelligent smart city architecture for vaccine manufacturing. *Computers & Industrial Engineering*, *166*, 107967.

Singh, S. K., & Park, J. H. (2022). TaLWaR: blockchain-based trust management scheme for smart enterprises with augmented intelligence. *IEEE Transactions on Industrial Informatics*, *19*(1), 626–634.

Singh, S. K., Park, J. H., Sharma, P. K., & Pan, Y. (2021). BIIoVT: Blockchain-based secure storage architecture for intelligent internet of vehicular things. *IEEE Consumer Electronics Magazine*, *11*(6), 75–82.

Solanki, S., & Solanki, A. D. (2020). Review of deployment of machine learning in blockchain methodology. *International Research Journal on Advanced Science Hub*, *2*(9), 14–20.

Srihith, I. V. D., Kumar, I. V. S., Varaprasad, R., Mohan, Y. R., Srinivas, T. A. S., & Sravanthi, Y. (2022). Future of smart cities: The role of machine learning and artificial intelligence. *South Asian Research Journal of Engineering and Technology*, *4*(5), 110–119.

Tanwar, S., Bhatia, Q., Patel, P., Kumari, A., Singh, P. K., & Hong, W. C. (2019). Machine learning adoption in blockchain-based smart applications: The challenges, and a way forward. *IEEE Access*, *8*, 474–488.

Teh, D., & Rana, T. (2023). The use of internet of things, big data analytics and artificial intelligence for attaining UN's SDGs. In *Handbook of Big Data and Analytics in Accounting and Auditing* (pp. 235–253). Springer Nature Singapore.

Thakur, N., Nagrath, P., Jain, R., Saini, D., Sharma, N., & Hemanth, D. J. (2021). Artificial intelligence techniques in smart cities surveillance using UAVs: A survey. *Machine Intelligence and Data Analytics for Sustainable Future Smart Cities*, 329–353.

Ullah, A., Qi, G., Hussain, S., Ullah, I., & Ali, Z. (2024). The role of LLMs in sustainable smart cities: Applications, challenges, and future directions. *arXiv preprint arXiv:2402.14596.*

Ullah, Z., Al-Turjman, F., Mostarda, L., & Gagliardi, R. (2020). Applications of artificial intelligence and machine learning in smart cities. *Computer Communications*, *154*, 313–323.

Ullah, Z., Al-Turjman, F., Moatasim, U., Mostarda, L., & Gagliardi, R. (2020). UAVs joint optimization problems and machine learning to improve the 5G and Beyond communication. *Computer Networks*, *182*, 107478.

Ulusar, U. D., Celik, G., & Al-Turjman, F. (2020). Cognitive RF-based localization for mission-critical applications in smart cities: An overview. *Computers & Electrical Engineering*, *87*, 106780.

Van Hoang, T. (2024). Impact of integrated artificial intelligence and internet of things technologies on smart city transformation. *Journal of Technical Education Science*, *19*(1), 64–73.

Waghmare, G., Kumar, S., Udhayakumar, R., Mannar, B. R., Singh, R., & Rawat, N. (2023, May). Utilizing social media platforms for artificial intelligence-based detection and developing effective communications in disaster-affected areas. In *2023 3rd International Conference on Advance Computing and Innovative Technologies in Engineering (ICACITE)* (pp. 501–504). IEEE.

Wei, Y., Yuan, H., & Li, H. (2024). Exploring the contribution of advanced systems in smart city development for the regeneration of urban industrial heritage. *Buildings*, *14*(3), 583.

Xia, Z., Du, J., Ren, Y., & Han, Z. (2022). Distributed artificial intelligence enabled aerial-ground networks: Architecture, technologies and challenges. *IEEE Access*, *10*, 105447–105457.

Yan, K., Liu, L., Xiang, Y., & Jin, Q. (2020). Guest editorial: AI and machine learning solution cyber intelligence technologies: new methodologies and applications. *IEEE Transactions on Industrial Informatics*, *16*(10), 6626–6631.

Yang, J., Xiao, S., Jiang, B., Song, H., Khan, S., & Ul Islam, S. (2020). Cache-enabled unmanned aerial vehicles for cooperative cognitive radio networks. *IEEE Wireless Communications*, *27*(2), 155–161.

Yu, H., & Zikria, Y. B. (2020). Cognitive radio networks for internet of things and wireless sensor networks. *Sensors*, *20*(18), 5288.

Zafeiropoulos, C., Tzortzis, I. N., Rallis, I., & Doulamis, A. (2023, March). Development of a support system for improved resilience and sustainable urban areas to cope with climate change and extreme events based on GEOSS and advanced modelling tools. In *International Conference on Transdisciplinary Multispectral Modeling and Cooperation for the Preservation of Cultural Heritage* (pp. 364–374). Springer Nature Switzerland.

Zahmatkesh, H., & Al-Turjman, F. (2020). Fog computing for sustainable smart cities in the IoT era: Caching techniques and enabling technologies-an overview. *Sustainable Cities and Society*, *59*, 102139.

6 Integrated Forensic Methodologies for Unmanned Aircraft Systems (UAS)

Devangana Sujay and Shishir Kumar Shandilya

6.1 INTRODUCTION

Quite interestingly, the term 'Drone' had not been considered the same throughout literature until technologists rose up to give it a touch of their own. A musical constant sound, a singular male counterpart within a bee colony to intelligent flying machines capable of cognitive skills must be above and beyond the rudimentary ideas revolving around this small terminology. Though in a thorough technical context, many a time, the term is used synonymously, a slight in its meaning is to be noticed.

6.1.1 UAS vs UAV vs Drone

Though the terms 'UAS', 'UAV', and 'Drone' are used interchangeably, there are differences among them.

The following table aims to clarify the distinguishing factors.

The journey of the evolution of this floating robot started off way before. A glance at it shall scale the intellectual ambitions of humans.

6.1.2 History of Drones

The pilotless aircraft had its rudimentary entry to serve the war. The first British radio-controlled aircraft was engineered in 1917, the following year, it witnessed the birth of its American rival. Contrastingly, neither of them were in action during World War 1. Though its earliest application was for military services, this broadened massively over the years, even within military uses such as its function of acting as a decoy, sending messages to men at the warzone, transporting types of equipment that it can carry, and so on and so forth. It is said that it is the British who came up with the terminology 'drone' (Imperial War Museums, 2024).

DOI: 10.1201/9781003481584-6

Factor	UAS	UAV	Drone
Definition	The entire system that connects the device that is airborne, the person and controller on the ground, and the link between both points.	The entity which is part of the UAS that is in the air, referring to the vehicle only.	The device that is committing its task in the air without a human pilot, but includes both the vehicle itself and whatever technology that powers its secure and meticulous execution of the task(s).
Example	A military surveillance system, a Vertical Take-Off and Landing or VTOL Drone (the aircraft), Visionair, a Ground Control Station or GCS software, and the link of communication between both.	A Freefly System Alta X, simply the vehicle and no other components included with it.	DJI Mini 4 Pro for aerial videography and photography, with the inclusion of a camera as a payload and/or sensors that may be integrated for additional functionalities.

FIGURE 6.1 Comparison among UAS, UAV, and drone

Karem, the founder of Karem Aircraft, was one of the earliest and the most credited personalities with respect to Drone Technology. His association with the US Defense Research Project Agency (DARPA) gave life to the Albatross—the first drone to be developed under this undertaking and of course, one of the most finely engineered ones of its times. Surprisingly, it weighed above 90 kg (~ 200 pounds) in its prototype stage. This has developed into what is now known as 'Predator Drones' (Drone Launch Academy, 2024).

With all that drone technology has traversed to become, at this current stage, something that cannot go unnoticed is the number of users who utilize this bright technology for nefarious activities. In fact, the count has only gone up in the past few decades throughout the world. The upcoming section explains the same using case studies. A recent report (July 2024–September 2024) by the Federal Aviation Administration came up with shocking statistics regarding the number of drone incidents in the United States, marking California with the highest count (Federal Aviation Administration [FAA], 2023). The violations include infiltration of physical or geographical privacy, for example, individual residences being watched using drones with cameras, no-fly zones, breach of laws in No-Fly Zone, such as near airports, military camps, and so on. Contraband activities are also common.[1] Most common drones are flown in the air, but drones such as amphibious drones that could go underwater also exist. Thus, airborne is the most common case that is used for a generic definition here, but what is implied is any environment that the vehicle may be in at any point of time in its path, including underwater or any other.

6.2 LITERATURE REVIEW

The primary obstacles in the forensic investigation of drones are the interpretation of flight data and the management of their multi-platform characteristics. The collection and analysis phases of the artifacts were conducted across several platforms, including UAS, mobile devices, and portable storage, showcasing a diverse array of file systems and interfaces. This mandates forensic tools to have customizable capabilities, requiring

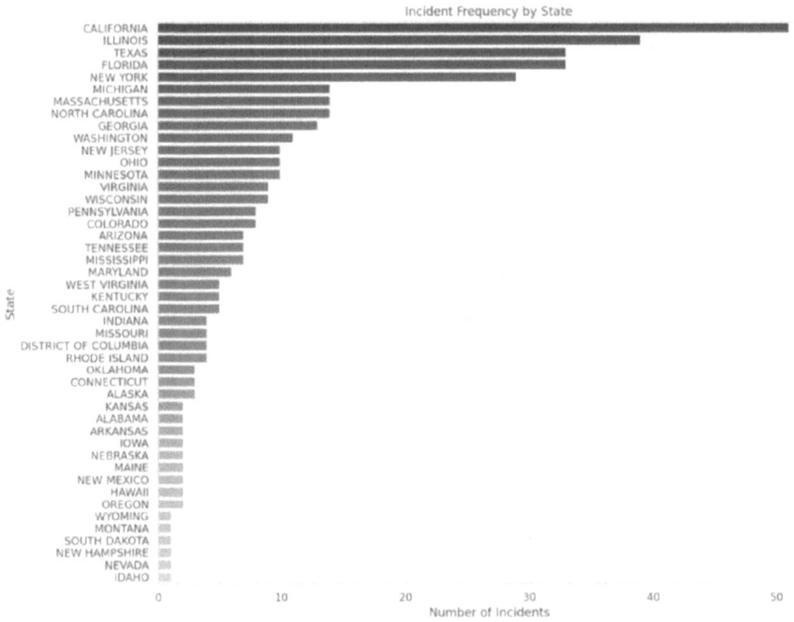

FIGURE 6.2 FAA's statistics on drone incidents in the United States

them to be compatible with numerous programming and/or scripting languages as well (Azhar, Barton, & Islam, 2018). The creation of a specialized tool for drone forensics timeline analysis is expounded upon by a study. The micro timeline on drones, especially the flight data, is not the most captivating feature of most tools. In order to create a more inclusive super chronology for the research, a blend of the macro and micro timeframes is the idea (Studiawan, Ahmad, Santoso, Shiddiqi, & Pratomo, 2022).

A refined picture of existing shortcomings of forensics with respect to drones is discussed elaborately in a research work by Bouafif, Kamoun, Iqbal, and Marrington (2018), which has a prime focus on methods, sources, and tools required to acquire forensically sound data from drones and paraphernalia. The characteristics of certain drones that utilize USB may create a hurdle to obtaining a forensic image. The sale (by vendor) and use (by customer) of non-standardized components of drones raise a significant conundrum. The Wipeout feature is a fairly mitigated area of discussion in this regard. This is mostly relevant if the human controller wishes to do so—a critical and probable anti-forensics practice.

In INTERPOL's Drone Countermeasure Exercise Report (INTERPOL, 2023), emphasis was given to C-UAS OR Counter-UAS, which had already gained the attention of numerous commercial suppliers as it proved to be the solution to meet a wide range of problems. The C-UAS systems can recognize, track, and detect the threat when a drone enters a designated monitored airspace. Nonetheless, drone countermeasures are a relatively recent technology that uses a number of automated methods to guarantee airspace security. Even though the potential advantages of C-UAS systems are receiving a greater spotlight, it is still challenging to benchmark these systems' capabilities. As a result, end users encounter difficulties matching the appropriate counter-drone solutions to the particular use cases. This is a key takeaway from the report.

6.3 SECURITY ISSUES RAISED BY DRONES

UAS usually relies heavily on network-based communication with the ground controller and the action undertaken by the drone thereafter. If this communication is not encrypted, interception of this line of contact can be easily performed. This could give rise to a spectrum of cybersecurity attacks, such as DDOS, to begin with. Owing to the varieties of sensors that may be integrated with the UAV and their respective functionalities along with their mode of communication and operation, a wider attack surface is opened up. Although the intent of using a drone and the components that are clubbed with it cannot be avoided, more and more security-enabling mechanisms are to be improvised depending on the usage of these specific components.

It was noted by a study that certain categories of drones lacked in their previous models, encryption capabilities when active flying mode is underway. The primary concern is the data manipulation that could be carried out at the phase, thus, a vulnerability and/or a potential threat. It was also noted that data may be sent to the cloud directly, especially sensor-related data; this could be a vulnerable dependency if not handled meticulously (Salamh et al., 2019).

6.3.1 CVE-2024-22520

This recent vulnerability exposure has circuitous implications for drone security. The National Vulnerability Database or NVD ranked it a base score of 8.2 (High) (National Vulnerability Database [NVD], 2024). It falls under the Common Weakness Enumeration (CWE) Name category: Authentication Bypass by Spoofing. What this means is that the attacker could act as a different person in control, better known as impersonation. It was discovered in Dronetag Drone Scanner 1.5.2. The target of this vulnerability processes certain HTTP requests. Given the case that a well-crafted HTTP request is sent, the probability of the malefactor being able to execute an arbitrary code is not meager. The patch for the same is disclosed. Since this is recent and a lack of awareness of the back-end functionalities and dependencies, such technical intricacies may be left untouched.

6.3.2 CVE-2023-47625

Another interesting CVE in this discipline is CVE-2023-47625 with its direct association with PX4 autopilot, a flight control solution. It has a base score of 4.3 (Medium) and is a Classic Buffer Overflow CWE. Remote control packets shall be capable of commanding over the working of the drone, leading it to malfunction or do task(s) that it was/were not intended to. Prior to 1.14.0, PX4-Autopilot versions were vulnerable to the flaw. This could just be the initial footsteps to Denial of Service or DOS attack and unauthorized gain of access. The overview of this category of attack could be thought like so: A home with an open door, once an invader gets in, access is obtained, what they wish to do next is beyond the security enabler's discretion (National Vulnerability Database [NVD], 2023a).

6.3.3 CVE-2023-6951

This particular vulnerability has a pinpointed target of DJI drone models. This flaw allows a hacker to obtain the WPA2 PSK key and access the drone's Wi-Fi network without authorization. This basically falls under the family of Use of Weak Credentials vulnerabilities. The vulnerability allows unauthorized access to the drone's communication system and can lead to data theft. It has a 6.6 (Medium) Base Score as per NVD (National Vulnerability Database [NVD], 2023b).

6.3.4 OWASP Top 10 Drone Security Risks

The following appeals to alterations that may be given attention to, particularly those pertaining to security mechanisms based on recognized security flaws from the web and mobile application domains for specialized drone environments (OWASP, 2023):

1. Insecure Communication Risk
2. Weak Authentication/Authorization Risk
3. Insecure Firmware/Software Risk
4. Inadequate Personal Data Protection Risk
5. Lack of Secure Update Mechanism Risk
6. Insecure Third-party Components Risk
7. Insufficient Network Security Risk
8. Physical Security Weaknesses Risk
9. Insecure Data Storage Risk
10. Lack of Logging and Monitoring Risk

With the widening of Drone-as-a-Service, these aforementioned security flaws are to shed insights and further hardening both from a security and forensics (pointing to investigative) perspectives.

6.4 CASE STUDIES ON DRONE INCIDENTS

The violations that were caused by these flying machines are beyond the bounds of predictions. It happens to be the same across the world with articles noting drones entering no-fly zones, sourcing aerial transport of illegal substances, and many more, popping up from various countries (Dedrone, 2024).

6.4.1 Cases in Europe

- Bordeaux National Police reported drone deliveries to prisoners which consisted of drugs and other objects. The investigation unearthed that the ground controller of the drone was a 17-year-old boy, the brother of one of the prisoners, among the other two contributors to the perpetration (Le Figaro, 2023).

- Germany reported repeated hovering of unidentified drones at its nuclear infrastructure near the coast of the North Sea, with further examination of the incident revealing that the breach had occurred previously as well, rendering the no-fly zone prone to illegal drone flights in recent times (Newsweek, 2024).
- Paris Olympics 2024 was also witness to numerous drone incidents, and it had been stated by officials that the Olympic sites had six drone interceptions per day on average. It was unconfirmed whether the drones belonged to and were maneuvered by curious tourists or other sources. Anti-drone defenses among other anti-aircraft defenses were put into operation to prevent any mishaps (Le Monde, 2024).

6.4.2 Cases in India

- At Amritsar Airport, three drones were detected in the airspace on August 24, 2024, which led to the disruption of flight operations and the further involvement of police and investigations thereafter. It had similar patterns to that compared with the incident at Imphal Airport after which the Directorate General of Civil Aviation declared premises and airports itself as 'Red Zones' (India Today, 2024). Another incident involved the Border Security Forces (BSF) successfully curbing an aerial transfer of heroin from cross-territory adversaries with the target of transporting the illegal material weighing above 6 kg over national boundaries at Amritsar. The mission was carried out in two phases, both of which were prevented by BSF. The canary set was noted to contain 560 gm, followed by a 5.570 kg package both attached to DJI Mavic 3 Classic drones (Web India 123, 2024).
- At Bengaluru, two flights had close encounters with unidentified drones setting chaos in the airport. Post-take on the first flight, it was reported to have been followed by a drone. This was instantly reported and later investigated by the airport police (CNBCTV18, 2024).

6.4.3 Cases in the USA

- The college football season was alarmed in Week 0 during the game of UMASS vs. New Mexico State when a miniature flying vehicle entered the stadium and soared low above the pitch. The game was brought to a short halt till the drone left the stadium (Boston.com, 2023).
- A dissimilar felony was committed by a Florida resident with his mission to transport US-sourced microelectronics outside US territory violating an array of laws. These microelectronics are capable of imparting military drones with specialized capabilities (DroneLife, 2024).

- Local County Sheriff's deputies had given prior intimation to residents regarding replacement and related works on the electric substation. This was operated using the Sheriff's Office UAS which was interrupted by or allegedly 'attacked' by the intruder at Vandalia, Ohio (ABC 17 News, 2023).

6.5 IMPORTANCE FOR FORENSICS IN UAS

Though an array of utilities are listed on the bright side, the picture of the solution to the all-important question 'What happens if something goes wrong to the mighty tech?' still seems way more bleak and vague than it should. As it is rightly stated by experts, the forensics techniques that are to back the technology itself are running at a slower pace than its technological advancements. The reason for the same branches out to be multiple, one of the core ones being differences in types of drones that are being manufactured and being onboarded into the market, with every new sensor, there is a new functionality, a whole new column added to the blueprint of data extraction from the same, tools used to make this possible, mutations to its integration, and so on and so forth.

What Law Enforcement Agencies or LEAs are aiming to do is to come up with a general strategy and methodology to forensically extract various sources of data related to UAS while keeping in mind the possibilities of waves of changes that may come through in the future.

6.6 STEPS FOR SEIZE AND SEIZURE FOR DRONES

At the crime scene or the environment of the incident, the geographic area needs to be secluded. Identification of the drone, its type, and the presence of any kind of vital source of physical evidence such as DNA, fingerprint, etc. is to be carefully analyzed and is among the first and foremost steps to be undertaken. If the camera lens is open, it needs to be closed. Doing this step is important so as to prevent the perpetrator from gathering any more information and for any unforeseen mishaps. In case of the presence of any other payload attached to the drone, a note of the same is to be made, as well as the physical state of the drone, whether damaged, and if so, specifics about the same.

LiPo or Lithium-Polymer batteries are mostly used alongside UAVs. It is to be dismantled carefully and needs to be transferred to a dry container. The following are to be photographed and recorded:

- Payload
- Damages, if any
- Make
- Model
- Serial number
- Federal Aviation Administration (FAA) number
- Federal Communications Commission (FCC) ID

Determination of the data storage type would come in handy for future steps. All artifacts are to be packaged, numbered, sealed, documented, and sent to a forensics lab for further investigation and analysis (Data Security Council of India [DSCI], 2023).

6.7 DRONE ARTIFACTS

Drone and mobile forensics have a binding to be broken while forensic examination is to be conducted which is now tied to a stronger grip with the wider application of encryption to flight data that are stored in mobile phones (both iOS and Android Phones).

An easily accessible artifact that can be acquired from the UAV (the hardware component) is the unique identification number or other serial number(s) embedded in it. Flight data must be the biggest point of focus when it comes to conducting forensics analysis in drones. This is mainly because of the capability of flight data to clarify flight activities. Images and videos may or may not prove useful, but the chances scale higher if plausible meta-data associated with the same could be presented in an admissible manner. Time stamps and GPS-related data are other vital sources (Salamh, Karabiyik, Rogers, & Al-Hazemi, 2019).

6.7.1 Drone Data Storage Sources

The following figure depicts the major sources from where a wealth of data related to a drone can be retrieved and the tools that can come in handy for the same. Please note that the efficiency of data recapture has a huge dependency on the make and model of the drone as well as the security mechanisms that may be implemented to add a layer(s) of protection or as per technical terminology, defense in depth. Additionally, specific sensors and/or other media related to the same, if present, are to be diverted attention to, in case deemed useful. As mentioned previously, for instance, encryption of flight data has been widely adopted; from an investigative standpoint, this would require the employed forensic tool to have such capabilities or be customizable enough for external plugins and/or be compatible with scripting. Thus, the given diagram may not be generalized for all categories of drones, especially at a point of time of innovation where accompanying constituents including hardware, software, and firmware are dynamically expanding.

6.7.1.1 GPS Modules

GPS Modules, to begin with, may not be present in each and every drone. But most commercial and professional drones are equipped with this, usually located on the body (top side) of the UAV underneath the GPS antenna, the most appropriate location so as to suit satellite visibility. The types of drones not integrated with a GPS module may be automated, which are self-aware of their home/base and destination coordinates as well as the path, otherwise, maybe a low-grade drone materialized to perform tasks not requiring and/or involving geographic locations.

Listed below are ways by which GPS modules contribute to drone forensics:

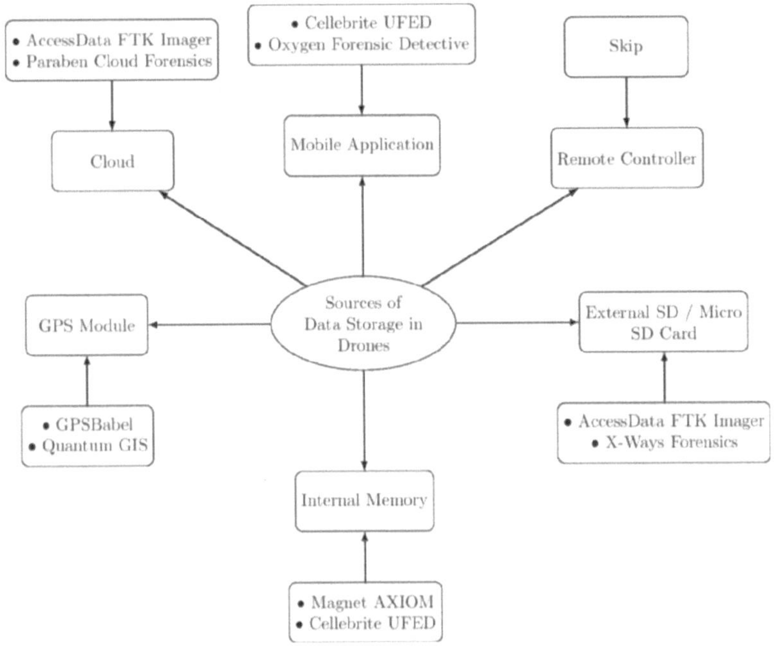

FIGURE 6.3 Sources of data storage in drones

- Flight Path Reconstruction: An array of vital particulars can be retrieved from flight logs which include but are not restricted to the following:
  - Waypoints
  - Home Coordinates
  - Destination Coordinates
  - Flight Path

The aforementioned details lay a blueprint of the flight undertaken by the drone, thus aiding investigators obtain a clearer picture of the incident.

- Geolocation Pointers: This would be the most precise and accurate detail that could be obtained as it capacitates the investigation to be directed to a pinpointed location. Geolocation coordinates are especially advantageous if the case at hand is related to trespassing or surveillance.
- Timestamp: All kinds of logs have this inbuilt feature. Its biggest merit is that it allows an investigator to chart out a very precise timeline of occurrences that may come in handy in confirming a hypothesis, for instance.
- Miscellaneous Sources: The following are a few key sources that could be paid attention to, if accessible:
  - Telemetry Logs
  - Camera Footage
  - Logs from other sensors, if applicable

The tools specified here are open-source tools. Owing to the fact that research is statistically less in this particular discipline of drone forensics and a more singled out substream of forensic data extraction from GPS Module, there are fewer tools available that are vouched for this specifically. Open-source tools are reliable in this context. Quantum GIS or QGIS is an abbreviation for quantum geographic information systems. Both tools help in obtaining GPS-related data by making them easy to understand in order to facilitate the process of deciphering specifics from these standards, which may be uneasy, if not represented comprehensively.

6.7.1.2 Internal Memory

- Flight Logs and Telemetry Data: It is an internal memory that houses crucial data such as detailed flight logs. This is inclusive of telemetry logs as well. Elements of interest here are listed as follows:
  - Altitude
  - Speed
  - Battery Details
  - GPS Coordinates
  - Timestamps
- Navigation Data: A more in-depth understanding can be gathered with the help of this class of evidence. The particulars here may include the following:
  - Waypoints
  - Geo-fencing
  - Return-to-Home Coordinates
- Firmware and Software Information: This information might seem to have a direct implication. Consider a scenario where the drone is susceptible to CVE-2023-47625, unauthorized changes may have been committed. Identifying this is best possible by checking if unauthorized firmware updates were preceded with, which is stored in the drone's operating system, and directly linked to the firmware and software of the flying machine.
- Error Logs: The following may be listed in an error log or any other diagnostic detail catalog:
  - System Failure
  - Lack of Signal
  - Malfunction

Even though the location(s) to be found out and forensically analyzed may be diverse, this primarily falls in the broader classification of memory forensics. So, tools that function best in the same are preferable, thus Magnet AXIOM and Cellebrit Universal Forensics Extraction Device (UFED).

6.7.1.3 External SD/Micro SD Card

- Metadata: If the drone is meant for aerial photography and/or videography, the corresponding photos and videos are most likely stored in the SD card, if available. The Exchangeable Image File Format (EXIF) data could be used, for instance, to confirm that the drone was present at X location at Y time, corroborating with the timeline analysis.

- Deleted Files: Irrespective of whether the deleted media is from footage, image, or log, it can be retrieved using standard forensic recovery tools. Deletion of files being a common anti-forensics practice, this capability proves useful from an investigation standpoint.

Most SD and micro SD cards utilize popular file systems like FAT 32 and exFAT. Please note that file system type is dependent on storage capability and compatibility factors as well. Since a file system may be required to forensically copy, forensics imaging may be carried out, thus the tool Forensic Toolkit, or FTK Imager. X-Ways Forensics also possesses comparable capabilities.

6.7.1.4 Remote Controller

If there exists the chance of redundancy in data collected from this source as well, it may be skipped. Remote controllers may directly communicate with other sources of data storage such as mobile applications, the cloud, etc. A circumstance when it is to be given consideration is if the remote controller is found in a compromised state suggestive of yielding useful information, be sure to proceed with the examination.

6.7.1.5 Mobile Application

The platform via which communication was linked with the drone evidently has the potential to render a chest of factual data vital to the forensic operation. (Many of the sources and factors relevant in this field are previously expounded, thus repetition is avoided here.)

- User Inputs and Control Commands: This is purely dependent on the feature of the app used. But some of them store the following data:
  - Commands issued during flight
  - Adjustment to settings
  - Changes committed during the operation

This could especially be proven handy in case the human controller's intent is to be analyzed. These details could prove what action the person was trying to undertake and what the motivation was for the same. It could be a factor to judge if legal parameters were kept intact.

- Configuration Settings: Configuration settings pertaining to the drone include the following:
  - Flight Modes
  - Camera Settings, if applicable
  - User Preferences

The drone's state at each mission can be analyzed from this data.

- User Profile and Accounts: For identification of the human operator this would be the most vital piece of evidence; a lot of ground control station platforms that are mobile applications utilize individual accounts for each session. This may also provide insights into user information and activity logs.
- Cache Files: Mobile applications tend to store volatile, operation-specific data that may include the most recent partial media or logs. This could be a key discovery.
- Communication Logs: Since the most relied upon communication line is between this platform and drone, alerts, notifications, and so on may also be present. Technical issues, if exist, can be clarified using the same.

Cellebrite UFED and Oxygen Forensic Detective are among the highest-ranked tools for mobile forensics.

The advantage of both of the tools is that they are compatible with iOS as well as Android devices.

6.7.1.6 Cloud

A fairly well-renowned mechanism for storing drone-related data is the utilization of a chosen cloud platform. If any other redundant physical storage unit is tampered with, this would remain an untouched source to check upon, even in a case without such an issue, cloud storage if applicable is a noteworthy point to conduct forensic analysis at. Real-time syncing of data is one of its attractive features for users, eliminating the need for hardware storage unit procurement. Historical data may also be stored here. A probable constraint here may be access control mechanisms. Dependency points to the kind of cloud environment used, so the selection of forensic tools to be employed must be carefully performed accordingly.

The best case scenario would be accessibility to all possible sources, obtaining forensically sound data and each detail corroborating the other positively, but clearly, that is not feasible in any case. Thus, attention is to be given to not omitting any vital source and for care to be taken in forensically acquiring necessary shreds of evidence as the admissibility of the same is to be taken into account.

6.8 FUTURE SCOPE

Automation of drones is a key lookout. This evidently shall be backed with numerous perks but inevitably its side of impediments. With the advent of AI regulations, the automation of drones shall be put under scrutiny; thus, research supporting its optimistic growth is predicted to find its limelight. The capabilities that drone technology is proclaimed to reach are unparalleled. The clubbing of AI has a touch of gold at this point of growth of tech. From a forensics stance, as more and more tools surface the market, their legitimacy is a critical concern. Privacy concerns are a technological as well a legal issue that drones cross over with. Policies are to step up in order to find solutions for the same. Meanwhile, the broadening capabilities of the already existing standardized tool is a curious way to counter complications in the view of practitioners. A deeper search into the division of drone artifacts would prove useful for understanding the requirements of the progressing forensic technology with respect to drones as well, for instance, a survey of commonly used sensors and the tools and methodologies to forensically extract evidence from them is an avenue of thorough research.

6.9 CONCLUSION

A clear picture of what drone technology implies to security, how security breach issues could be dealt with, its impact, and measures to surge through adversaries, from a security as well as forensic viewpoint, are discussed. While technology peaks new highs, its corresponding forensic capability must not diminish in caliber. At the phase of an incident, investigative insights are paramount; in a technologically sound context, the ability and the ease to conduct digital forensics processes while adhering to all mandated protocols shall transpire to be the need of the hour. Law enforcement agencies, academia, and crime-combating organizations are striving to attain this common goal of building resilience to UAS and ad-hoc technologies and also to have an eye ahead to what could rise in the years to come with the headway of this class of innovation.

REFERENCES

ABC 17 News. (2023, October 7). Vandalia man charged after allegedly trying to cause a drone crash. *ABC 17 News*. https://abc17news.com/news/crime/2023/11/08/vandalia-man-charged-after-allegedly-trying-to-cause-drone-crash/

Azhar, M., Barton, T. E. A., & Islam, T. (2018). Drone forensic analysis using open source tools. *Journal of Digital Forensics, Security and Law*, *13*(1), 6. https://doi.org/10.15394/jdfsl.2018.1559

Boston.com. (2023, October 7). UMass football game against New Mexico State delayed due to drone. *Boston.com*. https://www.boston.com/sports/sports-news/2023/08/26/umass-football-new-mexico-state-drone-delay-watch/

Bouafif, H., Kamoun, F., Iqbal, F., & Marrington, A. (2018). Drone forensics: Challenges and new insights. In *2018 9th IFIP International Conference on New Technologies, Mobility and Security (NTMS)* (pp. 1–6). IEEE. https://doi.org/10.1109/NTMS.2018.0010

CNBCTV18. (2024, October 7). Drone comes close to two Indigo flights at Bengaluru airport, probe ordered. *CNBCTV18*. https://www.cnbctv18.com/aviation/drone-comes-close-to-two-indigo-flights-at-bengaluru-airport-probe-ordered-17911761.htm

Data Security Council of India (DSCI). (2023). Aerial insights: The role of drone forensics in modern investigations. *Data Security Council of India*. https://www.dsci.in/resource/content/aerial-insights-role-drone-forensics-modern-investigations

Dedrone. (2024, October 7). Drone incident center. *Dedrone*. https://www.dedrone.com/resources/drone-incident-center

Drone Launch Academy. (2024). Who invented the drone? UAV history lesson. *Drone Launch Academy*. https://dronelaunchacademy.com/resources/who-invented-the-drone-uav-history-lesson/

DroneLife. (2024, October 7). Russian citizen arrested for illegally exporting drone technology to Russia. *DroneLife*. https://dronelife.com/2024/09/17/russian-citizen-arrested-for-illegally-exporting-drone-technology-to-russia/

Federal Aviation Administration (FAA). (2023). UAS sightings report. *Federal Aviation Administration*. https://www.faa.gov/uas/resources/public_records/uas_sightings_report

Imperial War Museums. (2024, October 7). A brief history of drones. *Imperial War Museums*. https://www.iwm.org.uk/history/a-brief-history-of-drones

India Today. (2024, August 28). Amritsar Sri Guru Ram Das Jee International Airport on high alert after drone detected in airspace. *India Today*. https://www.indiatoday.in/india/story/amritsar-sri-guru-ram-das-jee-international-airport-on-high-alert-after-drone-detected-in-airspace-2589333-2024-08-28

INTERPOL. (2023). Countering unmanned aerial systems (CUAS). *INTERPOL*. https://www.interpol.int/content/download/17737/file/CUASInterpol_Low_Final.pdf

Le Figaro. (2023, November 24). Prison de Gradignan: 5 individus arrêtés après des livraisons par drone. *Le Figaro*. https://www.lefigaro.fr/bordeaux/prison-de-gradignan-5-individus-arretes-apres-des-livraisons-par-drone

Le Monde. (2024, July 24). Paris 2024: Canadian football staff member detained over flying drone over rivals. *Le Monde*. https://www.lemonde.fr/en/sports/article/2024/07/24/paris-2024-canadian-football-staff-member-detained-over-flying-drone-over-rivals-6696648_9.html

National Vulnerability Database (NVD). (2023a). CVE-2023-47625. *National Institute of Standards and Technology*. https://nvd.nist.gov/vuln/detail/CVE-2023-47625

National Vulnerability Database (NVD). (2023b). CVE-2023-6951. *National Institute of Standards and Technology*. https://nvd.nist.gov/vuln/detail/CVE-2023-6951

National Vulnerability Database (NVD). (2024). CVE-2024-22520. *National Institute of Standards and Technology*. https://nvd.nist.gov/vuln/detail/CVE-2024-22520

Newsweek. (2024, October 7). Russia drones near Germany's NATO nuclear plant raise alarm. *Newsweek*. https://www.newsweek.com/russia-drones-germany-nato-nuclear-plant-1943384

OWASP. (2023). OWASP top 10 drone security risks. *OWASP*. https://owasp.org/www-project-top-10-drone-security-risks

Salamh, F. E., Karabiyik, U., Rogers, M., & Al-Hazemi, F. (2019). Drone disrupted denial of service attack (3DOS): Towards an incident response and forensic analysis of remotely piloted aerial systems (RPASS). In *2019 15th International Wireless Communications & Mobile Computing Conference (IWCMC)* (pp. 704–710). IEEE. https://doi.org/10.1109/IWCMC.2019.8766637

Studiawan, H., Ahmad, T., Santoso, B. J., Shiddiqi, A. M., & Pratomo, B. A. (2022). DroneTimeline: Forensic timeline analysis for drones. *SoftwareX, 20*, 101255. https://doi.org/10.1016/j.softx.2022.101255

Web India 123. (2024, June 29). News article on India. *Web India 123*. https://news.webindia123.com/news/Articles/India/20240629/4210236.html

7 Challenges and Risks of Drone Technology

Roheen Qamar and Baqar Ali Zardari

7.1 INTRODUCTION

The progress of globalization is being significantly impacted by technological integration. Drone technology is a noteworthy example of a technological invention. Applications for drones in a range of industries, including agriculture, medicine, and the military, have grown significantly in recent years. In agriculture, drones provide farmers with real-time data, allowing them to make informed decisions about the optimal use of farm inputs. Additionally, drones are important to the healthcare sector because they make it easier to transfer essential medical supplies—such as blood, vaccines, drugs, and laboratory tests—by air to remote locations in developing nations during emergency situations. Drones are used in the military to support targeted strikes and to track adversary movements, which improves security and surveillance operations. It is important to recognize that despite all of their advantages, drones can also present hazards, including the possibility of property damage and personal injury, particularly when used by inexperienced operators or in the event that a component fails while the drone is in flight. Extremist groups may be able to take control of drones and reroute the payload to suit their purposes (Emimi, Khaleel, & Alkrash, 2023).

Global Positioning System (GPS) technology and the availability of customizable software for smartphones and tablets have significantly increased flight durations, reliability, and user-friendliness. An illustration of a GPS spoofing attack is presented in Figure 7.1. The optimization of cameras and other sensors required for the effective use of drones in agriculture and natural resource management has also been made simpler by these advancements. The use of drones is now commonplace in many industries and has been engrained in the basic frameworks of both established and emerging economies. The consequences could be dire if control over this technology were to be lost, if not disastrous (Majeed, Abdullah, Mushtaq, & Kazmi, 2021).

In the medical field, drones are useful in numerous ways. These include transporting aid packages, medications, vaccines, blood, and other medical supplies to remote areas; ensuring that disease test samples and kits are transported securely in high-contagion areas; conducting assessments in disaster-affected areas where traditional access is severely restricted; and having the capacity to rapidly deploy automated external defibrillators to patients experiencing cardiac arrest, potentially saving lives during medical emergencies. It has been shown that drones can carry test kits, vaccinations, pharmaceuticals, personal protective equipment (PPE), and

DOI: 10.1201/9781003481584-7

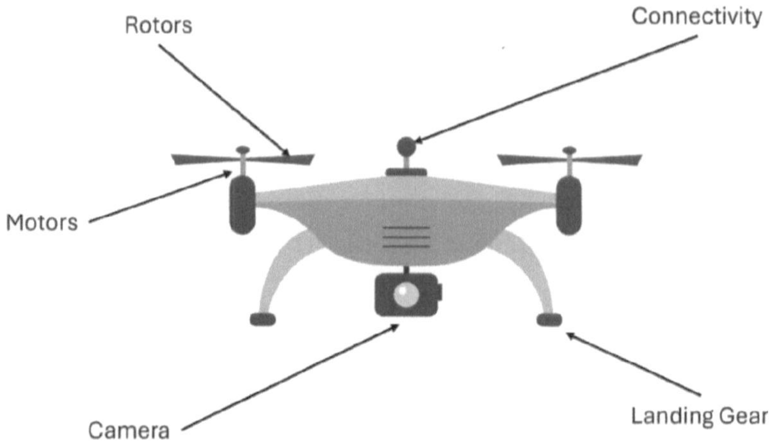

FIGURE 7.1 Architecture of drone

laboratory samples in the context of the COVID-19 pandemic. Drones may also help with automated inspections to make sure that social distancing rules are followed in public areas, which will help with management (Velev, Zlateva, Steshina, & Petukhov, 2019).

A new technology called drones offers customized solutions that are particularly helpful in cases of serious catastrophes, challenging terrain, and poor transportation infrastructure. The use of drones to swiftly and effectively deliver essential and life-saving medications to every member of society can significantly help achieve the goal of universal healthcare. Drone integration also presents prospects for increased efficiency in the transfer of passengers and commodities, which will help the logistics and transportation industries. Table 7.1 lists many drone kinds (Al-Dhaqm, Ikuesan, Kebande, Razak, & Ghabban, 2021).

To fully utilize this technology, regulatory concerns arising from the transition of drones from military to civilian applications must be addressed. Several studies have emphasized the need for clear legal frameworks that ensure the full potential of drones while simultaneously protecting people's rights to safety and privacy. The public is concerned about drone misuse, which could result in terrorist attacks, as well as concerns about invasions of privacy and military applications.

One of the main advantages of drones' numerous uses in the military, healthcare, and agricultural sectors is their ability to increase effectiveness, efficiency, and safety in a variety of operations and activities. The military's use of drones has changed war strategies due to their improved observation and reconnaissance capabilities. By facilitating real-time information gathering, target identification, and enemy movement surveillance, they provide armed forces with critical situational awareness. Because they reduce collateral damage and personnel hazards, drones are also crucial for accurate strikes. Drones allow the military to operate more precisely, cost-effectively, and responsively, which will ultimately improve mission success rates and lower casualties (Khalid et al., 2021).

In the medical field, drones have the potential to completely transform healthcare delivery, particularly in underserved and rural areas. They enable the prompt and efficient delivery of medical supplies, including prescription medications, immunizations, and emergency equipment, to far-flung areas. This is particularly useful in remote locations where it is challenging to use conventional modes of transportation or during emergencies like natural disasters.

In terms of improved farm management and more efficient farming methods, drones have a lot to offer the agricultural sector. Thanks to its real-time imagery, sensor data, and mapping capabilities, farmers can keep an eye on general field conditions, disease detection, and farmer wellness. This information enables focused interventions, precise input application, and data-driven decision-making on resource allocation. Due to their wide range of applications, drones have a significant impact on a number of fields, including improving data collection and analysis, increasing operational efficiency, and expanding access to necessary services. Drones provide precise and targeted operations, optimize resource utilization, and reduce risks to human life (Mohsan et al., 2023 - Labib et al., 2021).

7.2 APPLICATIONS OF DRONES

Drones have revolutionized military, medical, and agricultural operations, offering creative solutions. Advancements in drone technology have the potential to enhance efficiency, precision, and safety in several industries, addressing crucial obstacles and maximizing advantages.

7.2.1 Military Drones

Drone technological development has changed significantly over time, mirroring developments in contemporary Western military tactics. Drones are being used as vital assets in military operations as a result of this progress. Drones provide several benefits for the military, including the ability to save expenditures and lower

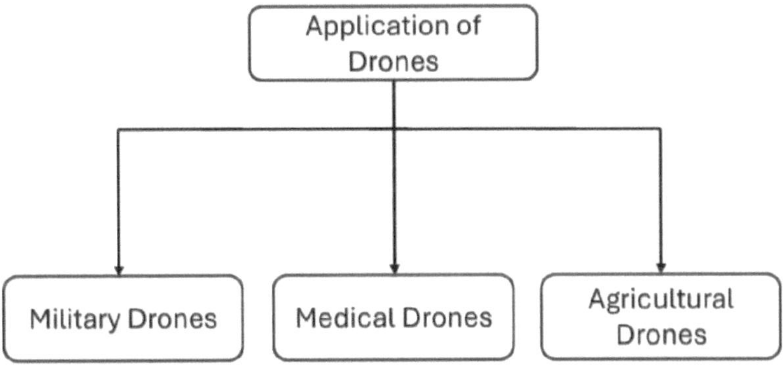

FIGURE 7.2 Application of drone

personnel hazards. Drones may be used for targeted murders and surveillance, which reduces the need for manned aircraft and the hazards to human pilots involved. Drones deployed for deadly surveillance have a tight connection to targeted assassinations in military operations. Decision-making processes involved in these interrelated systems have a direct impact on life-and-death decisions. Military forces may get intelligence in real time, carry out accurate monitoring, and launch targeted attacks with more accuracy and less collateral damage when they deploy drones for deadly surveillance (Rogers, 2021 - Park et al., 2021). It is important to remember, nevertheless, that using drones to carry out targeted executions presents difficult moral and legal questions. The assignment of life-and-death decisions to distant operators or automated systems presents concerns about transparency, accountability, and compliance with international rules governing armed combat.

In summary, the progress of rudimentary drone technology to its current application in contemporary Western warfare is demonstrated by its historical trajectory. Drones in the military have the potential to be advantageous in terms of cost and personnel danger, especially when it comes to targeted murders and deadly monitoring.

However, to guarantee the proper application and accountability in military operations, the employment of drones for these objectives requires serious ethical and legal considerations (Lee et al., 2023).

Drones are widely employed for intelligence collection, surveillance, and reconnaissance activities. Drones, equipped with modern sensors such as high-resolution cameras and infrared imaging, offer military forces real-time aerial photography, video feeds, and situational awareness. They can safely watch adversary movements, identify hazards, and acquire important information.

Drones are widely used in military applications because of their adaptability, agility, and operational benefits. Drones have several military applications, including:

- **Strikes Aimed Against Specific Targets:** Drones are used extensively in precise strikes and targeted operations. Armed drones, also known as unmanned combat aerial vehicles or armed drones, may carry and unleash precision-guided bombs. These systems let armed forces hit particular targets with precision and control, avoiding collateral damage and personnel dangers.

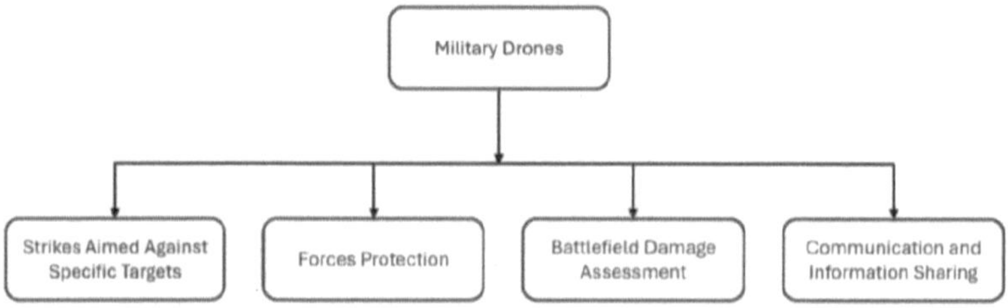

FIGURE 7.3 Military drones

- **Force Protection:** Drones are a great tool for enhancing force protection protocols. They can be employed to keep an eye on and protect military facilities, convoys, and sites. Armed forces operations can be made more secure by using drones equipped with sensors and surveillance capabilities to detect and track threats.
- **Battlefield Damage Assessment:** Drones can analyze the efficacy and impact of military operations and strikes. They may collect airborne pictures and data to assess operations, identify damage, and influence future decisions.
- **Communication and Information Sharing:** Drones can serve as communication relays in regions with inadequate or interrupted infrastructure. Drones can transmit data, sound, and video signals, allowing for continuous contact between military troops in remote or inaccessible areas (Purahong et al., 2022).

7.2.2 Medical Drones

Drones have become a key tool in the medical field, with the potential to change healthcare delivery through several uses. Some African nations, including Rwanda and Ghana, have had success implementing the helicopter delivery of medical supplies to health institutions in rural populations affected by poor road infrastructure and hilly terrain. Drones, according to a number of academic studies, provide underprivileged and remote areas with better medical care, faster response times, and lower transportation costs. As a reaction to the COVID-19 outbreak, the US government has granted permission to businesses like Zipline, among others, to transport food and medical supplies in certain areas. This has led to significant developments in the drone delivery industry (Truog et al., 2020).

According to recent academic research, the advances in drone technology in the environmental and ecological areas support the idea that drones can be extremely useful tools for public health, acting as skilled medical messengers. The utilization of medical drones is driven by their ability to provide precise delivery at a cost-effective rate when compared to traditional delivery methods. Emergency medicine empirical research has shown that it is safe and feasible to use drones to deliver automated external defibrillators (AEDs) in out-of-hospital cardiac arrest (OHCA) cases within areas identified by Geographic Information System (GIS) models (Quintanilla et al., 2021).

- **Aerial Delivery of Medical Supplies**: Drones provide fast and efficient delivery of medical supplies to remote areas. Drones can carry crucial medical supplies, like drugs, vaccinations, and blood samples, even in places with little road infrastructure and hard terrain. Drones have decreased response times, transportation costs, and healthcare access in underprivileged populations across Africa, including Rwanda and Ghana.
- **Immediate Medical Service**: Drones can significantly enhance emergency medical care. Drones equipped with AEDs can quickly respond to

cardiac arrests and provide emergency care before medical experts arrive. GIS models can help identify high-risk locations for cardiac crises, allowing drones to be strategically placed for faster reaction times and perhaps higher survival rates (Quintanilla et al., 2021).

- **Telemedicine Support**: As mobile communication hubs, drones can improve telemedicine. In remote or disaster-affected areas, drones equipped with contemporary communication technologies can establish wireless connectivity, enabling patients and medical professionals to have real-time video consultations. When physical access is restricted, this enables remote diagnosis, treatment advice, and control (Arunmozhiselvi et al., 2022).
- **Public Health Surveillance**: Public health surveillance can be aided by drones equipped with particular sensors and imaging systems. They are able to keep an eye on and identify environmental factors, including pollution, disease vectors, and air quality. Drones may collect real-time data and conduct aerial surveys to detect epidemics, track disease progress, and implement targeted treatments to reduce public health concerns (Bongomin et al., 2022).

7.2.3 Agricultural Drones

Agricultural drones, also known as agricultural unmanned aerial vehicles (UAVs) or agricultural unmanned aircraft systems (UAS), have emerged as ground-breaking tools in modern farming practices. These drones are outfitted and particularly intended to do a range of agricultural jobs.

Farmers and the agricultural sector as large can profit greatly from challenges. Crop monitoring, mapping, and analysis are just a few of the many tasks that agricultural drones can be used for. Agricultural drones that are outfitted with sophisticated imaging sensors, like thermal or multispectral cameras, are able to take high-resolution airborne photos, giving farmers the ability to track crop health, identify illnesses, and evaluate the general state of their farms. Improved agricultural yields and resource efficiency can result from using this data to inform decisions about pest control, fertilization, and irrigation (Arunmozhiselvi et al., 2022).

Drones are essential to precision agriculture because they make it easier to apply inputs like insecticides, fertilizers, and herbicides precisely. Agricultural drones can precisely and accurately target particular sections of a field, assuring optimum resource use and avoiding waste, by employing GPS and onboard sensors. This method, referred to as variable rate application, can enhance crop production while lowering costs and maintaining the environment. Agricultural drones help with land surveys and mapping in addition to crop management. Drones can provide precise 3D maps, digital elevation models, and pictures of agricultural landscapes by taking high-resolution photos and using photogrammetry techniques. Farmers may optimize field operations by using this data to help with land surveys, field boundary mapping, and the development of digital farm models (Bongomin et al., 2022).

Drones used for agriculture have shown to be successful in crop-spraying operations. These drones, which come with specialized spraying equipment, can administer pesticides, herbicides, and other agrochemicals to crops precisely. This has the benefit of reducing chemical consumption, applying the chemicals correctly, and increasing operational safety for farm personnel. The use of agricultural drones has enhanced the production of farming techniques, decreased expenses, and boosted efficiency.

These drones have the power to completely transform the agriculture sector by giving farmers fast, precise data, enabling precision farming methods, and streamlining resource management. However, for agricultural drones to be widely adopted and used sustainably, issues with legislation, privacy, and integrating drone technology into current agricultural systems must be resolved (Ayamga et al., 2021).

Agricultural drones, often known as agricultural UAVs, have several applications in current farming methods. Advanced aerial vehicles with sensors and imaging capabilities have altered agriculture, improving production, efficiency, and sustainability. Some important uses of agricultural drones include:

- Crop Monitoring and Health Assessment: Drones using high-resolution cameras and multispectral sensors may provide comprehensive images of crops. Specialized software analyzes images to detect crop health issues such as diseases, pests, nutritional deficits, and water stress. Early identification of concerns enables farmers to use precision agricultural strategies, optimize resource consumption, and increase crop yields.
- Precision Application of Inputs: Agricultural drones may apply fertilizers, insecticides, and herbicides to particular fields based on crop needs, decreasing misuse and environmental effects. This tailored application applies the proper quantity of inputs in the right area, resulting in cost savings and less chemical runoff.
- Field Mapping and Analysis: Drones using mapping software and GPS technology may provide precise 3D maps of fields, assisting farmers in assessing terrain, drainage patterns, and soil variability. This data helps develop site-specific management strategies and inform irrigation and drainage planning (Li et al., 2022).
- Crop Spraying: Drones with sprayers can efficiently and uniformly administer insecticides and crop protection goods across wide regions. Flying at low altitudes and following exact patterns provides consistent coverage, saving chemical waste and personnel expenses.
- Planting and Seeding: Some agriculture drones can carry seed dispensers for precise and efficient planting. This is particularly important for reseeding difficult-to-access regions or planting cover crops to enhance soil health.
- Livestock Monitoring: Drones using thermal cameras can monitor cattle on wide fields. This assists farmers in identifying and addressing health concerns, tracking lost livestock, and optimizing grazing practices.
- Irrigation Management: Drones may use thermal and multispectral photography to detect agricultural water stress levels and inform irrigation management decisions. Identifying locations with water stress allows farmers to change irrigation schedules and practices, improving water conservation and efficiency (Keshet, Brook, Malkinson, Izhaki, & Charter, 2022).

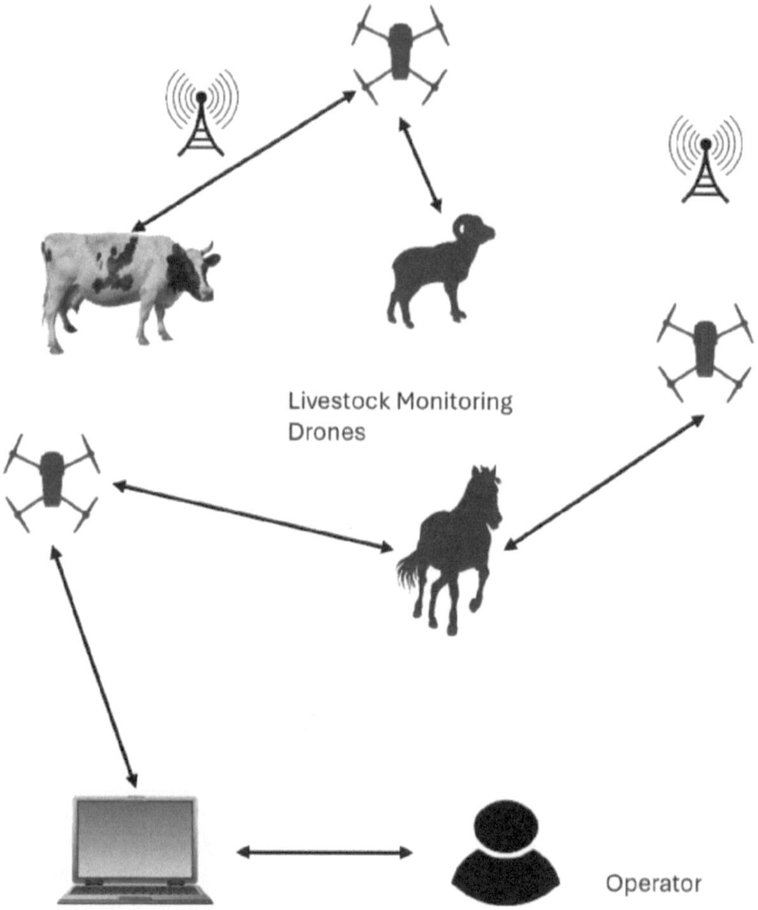

FIGURE 7.4 Live stoke drone monitoring

7.3 OPPORTUNITIES AND CHALLENGES

7.3.1 Opportunities and Challenges of Military Drones

Military drones offer advantages like surveillance and precision strikes, but they also pose challenges such as ethical concerns, vulnerability to cyberattacks, and potential misuse. This allows for proactive decision-making, target identification, and tracking of adversary movements.

- Precision Attacks: Military drones allow for accurate and targeted attacks, limiting collateral damage and personnel danger. They enable operations in complicated and high-risk areas, resulting in mission accomplishment with minimum human engagement. Military drones can dramatically cut expenses and dangers compared to human aircraft operations. Drones reduce the need for human pilots, lowering risk and perhaps saving lives.

- Regulatory and Legal Frameworks: The deployment of military drones presents complicated ethical, legal, and international law concerns. Drone regulations, such as airspace limitations, rules of engagement, and privacy issues, require careful consideration to guarantee responsible and accountable use.
- Cyber Security Threats: Military drones are vulnerable to hacks, compromising their operational integrity and data security. Drone systems require strong cyber security protections to prevent unwanted access or control.
- Public Perception and Acceptance: The deployment of military drones raises worries about civilian fatalities, privacy infringement, and autonomous decision-making. Transparent communication and participation with the public are essential for establishing trust and addressing misunderstandings (Kaya, Goraj, & Institute of Aeronautics and Applied Mechanics, 2020).

7.3.2 Opportunities and Challenges of Medical Drones

- Rapid and Remote Healthcare Delivery: Medical drones provide medical supplies, such as vaccinations, drugs, and emergency equipment, quickly and efficiently to distant or inaccessible locations. They increase access to healthcare in neglected areas, especially during crises and natural disasters.
- Telemedicine Support: Medical drones can improve telemedicine by allowing remote consultations, real-time video feeds, and communication between healthcare experts and patients in remote areas.
- Emergency Medical Response: Medical drones can bring life-saving equipment, such as automated external defibrillators (AEDs), to emergency scenes, leading to faster reaction times and better survival rates for cardiac arrest.
- Payload Capacity and Range: Medical drones are limited in their ability to transport bigger payloads, such as medical supplies or equipment. Improvements in payload capacity, battery life, and range are necessary to optimize their Effectiveness.
- Infrastructure and Logistics: Establishing infrastructure, including landing pads and charging stations, is vital for the efficient deployment of medical drones. Addressing logistical issues, such as inclement weather and airspace cooperation, is crucial for successful operations (Purahong et al., 2022).

7.3.3 Opportunities and Challenges of Agricultural Drones

- Precision Agriculture: Agricultural drones provide precision monitoring and data collecting, helping farmers optimize resource allocation, identify crop diseases, and improve farming operations. They enhance precision agricultural practices, increasing crop yields and resource efficiency.
- Crop Health and Management: Agricultural drones use powerful image sensors to deliver vital insights about crop health, nutrient deficits, and insect infestations. Data-driven decisions enable farmers to make timely interventions and allocate resources more effectively.

- Mapping and Analysis: Agricultural drones provide precise field maps, digital elevation models, and orthomosaic photos for land surveys, irrigation planning, and border mapping. This data helps improve farm management and decision-making processes.
- Data Analysis and Interpretation: Analyzing enormous amounts of data collected by agricultural drones can be challenging and time-consuming. To gain relevant insights from drone-collected data, farmers want easy-to-use software and analytical tools.
- Regulations and Safety: Agricultural drone activities must follow aviation rules and safety standards. Compliance with airspace rules, privacy concerns, and minimizing dangers when flying near inhabited areas are crucial difficulties to overcome.
- Cost and Scalability: Small-scale farmers may have challenges in procuring and maintaining agricultural drones due to their high cost. The scalability and affordability of drone technology, including equipment, training, and support services, are crucial for increasing acceptance and accessibility (Hafeez et al., 2023).

7.4 ADVANCEMENTS IN DRONE SENSOR TECHNOLOGY

Drone technology has become a disruptive force that is changing several sectors and redefining how information is gathered and analyzed and how decisions are made. Drone sensor technology, a vital component that enables unmanned aerial vehicles (UAVs) to sense their environment and collect vital data, is at the core of this game-changing capability. With their wide range of sensors, drones can take pictures, measure distances, identify changes in their surroundings, and navigate over challenging terrain on their own. These sensors are essential for improving flight control, navigation, and data collection for a variety of uses, including environmental monitoring, infrastructure inspection, disaster relief, and precision agriculture. An overview of the main drone sensor technologies, their features, and their numerous applications across a range of sectors are given in this brief introduction (Park, Seo, Om, & Kim, 2020).

It looks at how important drone sensor technology is to open up new UAV applications and lay the groundwork for a more intelligent and connected future. Improvements in sensor technology are expected to spur further innovation and increase the potential of drones in many domains as drone technology develops, influencing the course of aerial intelligence and revolutionizing global enterprises. Here are some additional specifics on some of the major drone sensor technologies:

- GPS (Global Positioning System) Sensor: This vital sensor allows the drone to have exact navigation and positioning while in flight by giving it precise location data. The drone's GPS determines its latitude, longitude, and altitude by receiving signals from many satellites. This allows it to fly along predefined routes or hold steady locations when hovering.

- Sensors for Imaging: Drones are frequently outfitted with diverse imaging sensors to obtain visual information and pictures from the air. While multispectral cameras may record data from several small regions of the electromagnetic spectrum, RGB cameras only record regular color pictures. This makes it possible to analyze the qualities of the soil, crop stress, and vegetation health.
- Light Detection and Ranging, or Lidar, Sensor: Lidar sensors pulse out laser light and time how long it takes for the light to return to the drone. High-accuracy 3D maps of the landscape, structures, and other ground-based objects are produced using this data (Karachalios, Kanellopoulos, & Lazarinis, 2021). Applications including topographical mapping, forestry management, urban planning, and infrastructure inspection benefit greatly from the usage of lidar.
- Unit of Inertial Measurement (IMU): The orientation, acceleration, and angular velocity of the drone are measured by the IMU, which combines gyroscopes and accelerometers. The flight controller can provide smooth flight and control by continuously checking these parameters, which allows it to steady the drone and modify its location in real time.
- Ultrasonic Indicators: When a sound wave strikes something beneath the drone, ultrasonic sensors record how long it takes for the sound to return. In low-altitude settings, this data permits object recognition and avoidance and aids in the drone's ability to maintain a steady height during flight (Khuzaimah et al., 2022).
- Barometer: Changes in altitude are closely correlated with changes in air pressure, which is measured by the barometer. This sensor is crucial for keeping the aircraft at a constant height, particularly in areas where the atmospheric pressure varies.
- Magnetometer: The magnetometer detects changes in the Earth's magnetic field, providing data on the drone's heading and orientation relative to the Earth's magnetic north. This information is used for accurate navigation and maintaining proper flight direction.
- Chemical and Gas Sensors: Certain drones are designed to monitor the environment and identify certain gasses or dangerous materials. These drones come with chemical and gas sensors installed. Applications such as industrial inspections, emergency response, and air quality monitoring depend heavily on these sensors (Gerwen et al., 2022).

7.5 DRONE COMMUNICATION TYPES

Four main categories apply to the communications that drones use: Direct communication between two or more drones is known as drone-to-drone (D2D); communication between drones and ground-based stations is known as drone-to-ground station (D2GS); communication between drones and network infrastructure is known as drone-to-network (D2N); and communication between drones and satellites in orbit is known as drone-to-satellite (D2S). In order to provide smooth and effective data interchange, control, and coordination in a variety of drone operations and applications, several communication types are essential.

Figure 7.5 illustrates the different possible drone communication.

- **Drone-to-Network:** Communication makes it possible to choose an appropriate network according to the required degree of security. Cellular communication using frequencies like 3 GHz, 4 GHz, 4G+ (LTE), and 5 GHz may be a part of this communication type. To protect the integrity and confidentiality of these wireless communication networks, it is crucial to ensure their security.
- **Drone-to-Satellite:** Communication is crucial for delivering real-time coordinates using the GPS. This feature allows drones to be recalled to their original location if they leave the line of sight or beyond the control range. Satellite communications are very secure and dependable, making them suitable for crucial applications. Satellite communications are mostly used by armed forces and specialized operations due to their high price and maintenance requirements. Satellite-based communication systems provide

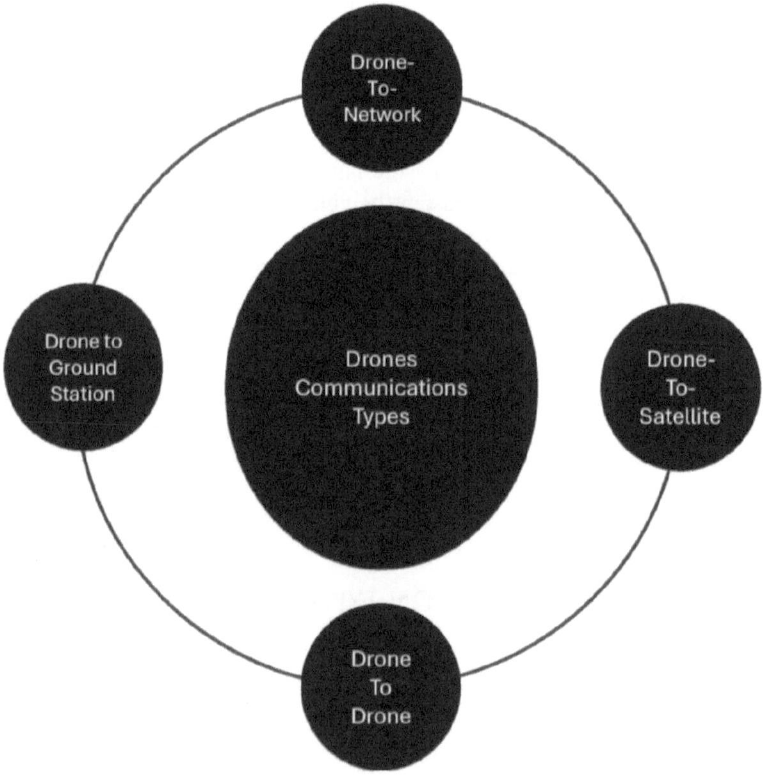

FIGURE 7.5 Drone communication types

advantages, but their adoption needs careful consideration of finance and operational requirements.

- **Drone to Drone:** Communication is not yet standardized and is under active investigation. Machine Learning can optimize an intelligent UAV-based wireless communication system, as indicated in the reference. Drone-to-drone (D2D) communication is often referred to as peer-to-peer (P2P). However, the P2P paradigm makes communication vulnerable to attacks like jamming, D-DoS, and Sybil.
- To ensure the durability of Drone-to-Drone communication systems against possible attacks, it's necessary to solve security concerns throughout implementation and advancement.
- **Drone-to-Ground Station:** Communication follows defined industrial protocols that use wireless technologies such as Bluetooth and Wi-Fi 802.11, operating at 2.4 GHz and 5 GHz frequencies. Most drone-to-ground communications are public and lack strong security precautions. Single-factor authentication can be subject to breaches, allowing both passive eavesdropping and aggressive man-in-the-middle attacks. Improving drone-to-ground communication security is crucial for protecting sensitive data and preventing illegal access. Implementing better authentication and encryption approaches can reduce vulnerabilities and protect communication between drones and base stations (Sharma et al., 2020).

7.6 LITERATURE REVIEW

Boccadoro, Striccoli, & Grieco, 2021: The work aims to categorize the multifaceted aspects of the Internet of Things (IoT) by focusing on its applications, applicability, and economic/social implications. It follows the structure of the Internet protocol stack, examining cross-layer and optimization approaches. The study also examines privacy and security considerations and discusses research challenges and future directions. The work also highlights open issues and promising technologies for further development in the Internet of Drones (IoD) field.

Ayamga et al., 2021: Drone technology is increasingly used in agriculture, health, and military sectors for real-time data, medical supplies, and security. However, it can also cause injuries and damage if not trained and can be hijacked by extremists. This paper presents a Strengths, Weaknesses, Opportunities, and Threats analysis of drone developments in these fields.

Yaacoub, Noura, Salman, & Chehab, 2020: This survey investigates the threats of drones in cyber-attacks and their potential countermeasures. It reviews the various uses of drones for malicious purposes, including communication links and smart devices. The paper presents a detailed review of drone usage in civilian, military, and terrorism domains and presents a realistic attack scenario. This review aids ethical hackers in understanding vulnerabilities and developing new techniques for enhanced UAV attack detection and protection.

Rejeb, Abdollahi, Rejeb, & Treiblmaier, 2022: Drones have revolutionized agriculture by offering cost savings, increased efficiency, and improved profitability. A comprehensive review of academic literature on agricultural drones reveals critical topics such as remote sensing, precision agriculture, deep learning, machine learning, and the Internet of Things. The study suggests six research clusters and suggests future directions in drone research in agriculture.

Park et al., 2021: The authors in this paper surveyed anti-drone systems, focusing on non-military drone usage and analyzing detection, identification, and neutralization technologies. It proposes a hypothetical system for adaptable defense, discusses potential drone-side safety schemes, and proposes future solutions to address challenges in constructing military-grade anti-drone systems due to installation and operation costs.

7.7 POSSIBLE FUTURE WORKS IN DRONES

As we approach a new era, it is apparent that the drone scene has changed dramatically. What was once a budding business is now set to challenge global conventions. We are not just seeing change; we are also participating in a drone revolution.

A brief overview of drone technology growth reveals remarkable improvements. Drones have progressed from military uses to consumer households, business operations, and even possibly life-saving medical devices.

Drone cameras are transforming aerial photography as technology advances. Drone photography is no longer limited to high-end projects; it is revolutionizing social media landscapes and amateur filmmaking.

The drone business is also targeting a younger market. In kid-friendly drones, education and fun are perfectly integrated. They're more than simply toys; they're instruments that promote learning and creativity.

For those who value security, the sky's watchful eye serves as a personal protector. Personal security drones with cameras and warning systems provide peace of mind and an additional layer of home security.

Drones have become the new lifelines in remote locations, transporting critical medical supplies. Drones guarantee that medicines and vaccinations reach every part of the planet.

In the drone world, little is the new large. Nano drones, which are small yet strong, have potential applications in surveillance, research, and perhaps pollination.

Drones are becoming commonplace. Drones will be used for everything from delivery to photography, security, and healthcare in the future. As we embrace the drone age, we must do so wisely. With each revolution comes the obligation to manage change. The future of drones beckons, with hopeful trends and forecasts that will alter the world as we know it (Sindiramutty et al., 2024).

7.8 CONCLUSION

Drones provide real-time images and sensor data from agricultural areas, allowing farmers to make educated decisions about farm inputs. They are an effective way to bring medical supplies to isolated places where road infrastructure is limited by tough geography, making traditional modes of transportation unfeasible. Using airborne drones can improve medical workers' efficacy and efficiency, perhaps saving more lives. The military may use drones responsibly while protecting civilians.

Drone technology improvement relies heavily on research and development, which may solve vulnerabilities and dangers in several fields. Establishing cross-border legislation is crucial for properly using drone technology. To prevent unlawful use of drones, users and future users should be informed of current rules. Existing transportation and supply linkages need to be better understood, as integrating drones with current transportation infrastructure and supply chains requires further research. To promote drone adoption in underdeveloped nations, it is important to expand payload capacity and flight durations while also taking cultural factors into account.

Drone uses in agriculture, healthcare, and the military require a holistic strategy that includes research, regulatory frameworks, and integration with current systems. Embracing these activities can open up avenues to the full potential of drone technology, resulting in increased efficiency, effectiveness, and acceptability in specific industries and globally.

REFERENCES

Al-Dhaqm, A., Ikuesan, R. A., Kebande, V. R., Razak, S., & Ghabban, F. M. (2021). Research challenges and opportunities in drone forensics models. *Electronics*, *10*(13), 1519.

Arunmozhiselvi, S., Ananth Kumar, T., Manju Bala, P., Usharani, S., & Glorindal, G. (2022). A systematic approach to agricultural drones using a machine learning model. In *Machine Learning Approaches and Applications in Applied Intelligence for Healthcare Data Analytics*, Boca Raton: CRC Press, pp. 41–60.

Ayamga, M., Akaba, S., & Nyaaba, A. A. (2021). Multifaceted applicability of drones: A review. *Technological Forecasting and Social Change*, *167*, 120677.

Boccadoro, P., Striccoli, D., & Grieco, L. A. (2021). An extensive survey on the Internet of Drones. *Ad Hoc Networks*, *122*, 102600.

Bongomin, O. et al. (2022). Applications of drones and image analytics in field phenotyping: A potential breakthrough in Uganda's agricultural research. *SSRN Electronic Journal*. https://doi.org/10.2139/ssrn.4158755

Emimi, M., Khaleel, M., & Alkrash, A. (2023). The current opportunities and challenges in drone technology. *International Journal of Electrical Engineering and Sustainability*, *1*, 74–89.

Gerwen, J. V.-V., Geebelen, K., Wan, J., Joseph, W., Hoebeke, J., & De Poorter, E. (2022). Indoor drone positioning: Accuracy and cost trade-off for sensor fusion. *IEEE Transactions on Vehicular Technology*, *71*(1), 961–974.

Hafeez, A. et al. (2023). Implementation of drone technology for farm monitoring & pesticide spraying: A review. *Information Processing in Agriculture*, *10*(2), 192–203.

Karachalios, T., Kanellopoulos, D., & Lazarinis, F. (2021). Arduino sensor integrated drone for weather indices: A prototype for pre-flight preparation. *JITA - Journal of Information Technology Case and Application (Banja Luka) - APEIRON*, *21*(1), 2021.

Kaya, S. (2020). Institute of Aeronautics and Applied Mechanics - Warsaw University of Technology, Z. Goraj, and Institute of Aeronautics and Applied Mechanics, The use of drones in agrıcultural production. *International Journal of Innovative Approaches in Agricultural Research*, *4*(2), 166–176.

Keshet, D., Brook, A., Malkinson, D., Izhaki, I., & Charter, M. (2022). The use of drones to determine rodent location and damage in agricultural crops. *Drones*, *6*(12), 396.

Khalid, M., Namian, M., & Massarra, C. (2021). The dark side of the drones: A review of emerging safety implications in construction. *EPiC Series in Built Environment*, *2*, 18–27.

Khuzaimah, Z., Nawi, N. M., Adam, S. N., Kalantar, B., Emeka, O. J., & Ueda, N. (2022). Application and potential of drone technology in oil palm plantation: Potential and limitations. *Journal of Sensor*, *2022*, 1–18.

Labib, N. S., Brust, M. R., Danoy, G., & Bouvry, P. (2021). The rise of drones in internet of things: A survey on the evolution, prospects and challenges of unmanned aerial vehicles. *IEEE Access*, *9*, 115466–115487.

Lee, J., & Kwon, S. (2023)."A basic study on the selection of required operational capability for attack drones of Army TIGER units using AHP technique. *Journal of the Korea Institute of Military Science and Technology*, *26*(2), 197–204.

Li, J. et al. (2022). Rapid evaluation model of endurance performance and its application for agricultural UAVs. *Drones*, *6*(8), 186.

Majeed, R., Abdullah, N. A., Mushtaq, M. F., & Kazmi, R. (2021). Drone security: Issues and challenges. *Parameters*, *2*(5). https://doi.org/10.14569/IJACSA

Mohsan, S. A. H., Othman, N. Q. H., Li, Y., Alsharif, M. H., & Khan, M. A. (2023). Unmanned aerial vehicles (UAVs): Practical aspects, applications, open challenges, security issues, and future trends. *Intelligent Service Robotics*, *16*(1), 109–137.

Park, J.-H., Seo, Y.-G., Om, S.-Y., & Kim, Y.-H. (2020). Implementation of drone system for measuring multilane traffic information based on laser sensor. *The Journal of Korean Institute of Information Technology*, *18*(12), 113–119.

Park, S., Kim, H. T., Lee, S., Joo, H., & Kim, H. (2021). Survey on anti-drone systems: Components, designs, and challenges. *IEEE Access*, *9*, 42635–42659.

Purahong, B., Anuwongpinit, T., Juhong, A., Kanjanasurat, I., & Pintaviooj, C. (2022). Medical drone managing system for automated external defibrillator delivery service. *Drones*, *6*(4), 93.

Quintanilla García, I., Vera Vélez, N., Alcaraz Martínez, P., Vidal Ull, J., & Fernández Gallo, B. (2021). A quickly deployed and UAS-based logistics network for delivery of critical medical goods during healthcare system stress periods: A real use case in Valencia (Spain). *Drones*, *5*(1), 13.

Rejeb, A., Abdollahi, A., Rejeb, K., & Treiblmaier, H. (2022). Drones in agriculture: A review and bibliometric analysis. *Computers and Electronics in Agriculture*, *198*, 107017.

Rogers, J. (2021). Future threats: Military UAS, terrorist drones, and the dangers of the second drone age. In *Drone warfare: Trends and emerging issues* (pp. 481–505). The Joint Air Power Competence Centre.

Sharma, A., Vanjani, P., Paliwal, N., Basnayaka, C. M. W., Jayakody, D. N. K., Wang, H. C., & Muthuchidambaranathan, P. (2020). Communication and networking technologies for UAVs: A survey. *Journal of Network and Computer Applications*, *168*, 102739.

Sindiramutty, S. R., Jhanjhi, N. Z., Tan, C. E., Khan, N. A., Shah, B., Yun, K. J., & Hussain, M. (2024). Future trends and emerging threats in drone cybersecurity. In *Cybersecurity Issues and Challenges in the Drone Industry* (pp. 148–195). IGI Global.

Truog, S. et al. (2020). Insights before flights: How community perceptions can make or break medical drone deliveries. *Drones*, *4*(3), 51.

Velev, D., Zlateva, P., Steshina, L., & Petukhov, I. (2019). Challenges of using drones and virtual/augmented reality for disaster risk management. *The International Archives of the Photogrammetry, Remote Sensing and Spatial Information Sciences*, *42*, 437–440.

Yaacoub, J. P., Noura, H., Salman, O., & Chehab, A. (2020). Security analysis of drones systems: Attacks, limitations, and recommendations. *Internet of Things*, *11*, 100218.

Index